Water Hydraulics Control Technology

Erik Trostmann
Technical University of Denmark
Lyngby, Denmark

CRC Press
Taylor & Francis Group
Boca Raton London New York

CRC Press is an imprint of the
Taylor & Francis Group, an **informa** business

CRC Press
Taylor & Francis Group
6000 Broken Sound Parkway NW, Suite 300
Boca Raton, FL 33487-2742

First issued in paperback 2019

ISBN-13: 978-0-8247-9680-8 (hbk)
ISBN-13: 978-0-367-40154-2 (pbk)

This book contains information obtained from authentic and highly regarded sources. Reasonable efforts have been made to publish reliable data and information, but the author and publisher cannot assume responsibility for the validity of all materials or the consequences of their use. The authors and publishers have attempted to trace the copyright holders of all material reproduced in this publication and apologize to copyright holders if permission to publish in this form has not been obtained. If any copyright material has not been acknowledged please write and let us know so we may rectify in any future reprint.

Library of Congress Cataloging-in-Publication Data

Trostmann, Erik.
    Water hydraulics control technology / Erik Trostmann.
      p. cm.
    Includes bibliographic references and index.
    ISBN 0-8247-9680-2 (acid-free paper)
    1. Hydraulic control. 2. Hydraulics. I. Title.
TJ840.T77 1996
629.8'042—dc20
                                    95-31148
                                    CIP

Visit the Taylor & Francis Web site at
http://www.taylorandfrancis.com

and the CRC Press Web site at
http://www.crcpress.com

Water Hydraulics
Control Technology

FOREWORD

In 1961 Danfoss started manufacturing hydraulics for low-speed, high-torque motors and hydrostatic steering units on a license from Char-Lynn, Minneapolis, USA. Since then, Danfoss has become one of the largest producers of these innovative products in the world.

Over the years many other innovative products have been developed, and Danfoss has more or less built a tradition to produce innovative products.

One day in 1989 Vagn Bender, then President of the Hydraulics Division, came over to my office and said that he thought Danfoss should develop water hydraulics. At that time I was head of our corporate research department. We both thought it was a great challenge, but at the same time we were very skeptical as to our chances of succeeding in such a demanding development; in fact we both laughed a little. Nevertheless, we decided to go ahead anyway.

The whole development project lasted 5 years in total secrecy. It is a long story in itself and is a model case for how to utilize all the resources of our team and to stand on the shoulders of all that was known from literature, patents, and the best experts in the world.

After only 6 months we knew that it could be done. The remaining time was used to develop products and make service life tests.

Such a success story naturally has many fathers, but I would particularly like to mention Vagn Bender. Not only did he supply the vision, but his leadership, determination, willpower, and judgement were instrumental in making this project a success.

It is our intention with this book to share the many advantages of water hydraulics, which we believe will develop into an entirely new branch within actuator technology, as it will open up new and attractive possibilities within many application areas.

Jørgen M. Clausen
President, Mobile Hydraulics Division
Danfoss

INTRODUCTION

For a long time, water was the only pressure medium considered for hydraulics—its use was endorsed by Joseph Bramah, the inventor of the hydraulic press, as far back as 1795.

At the beginning of the twentieth century, the use of mineral oil instead of water led to significant technical improvements because of its higher viscosity and superior lubricating properties.

Since then, almost a century of intensive technical development has passed. However, it has become manifest that today's high degree of technical progress has inflicted harmful effects on the environment. Increasing public awareness of environmental issues calls for measures designed to prevent ecological damage to the greatest extent possible.

It is at this point that engineers should begin to reconsider pure water as an inherently environmentally friendly pressure medium for hydraulics.

Today, technical processes and modern materials are available to the engineer, which allow hydraulic components to be adapted for use with pure tap water. Today's engineers are challenged to bring water hydraulics up to the same high standard achieved by oil hydraulics, which has developed over the course of many decades.

Although a number of companies offer water hydraulic components and systems, the advantages and disadvantages of this particular drive technology have hitherto been known to only a handful of specialists.

The author of this book, Professor Erik Trostmann, deserves praise for presenting an objective and comprehensive survey of water hydraulics—in terms of not only the pressure medium itself, but also the necessary components and systems—with all its advantages and disadvantages. Its advantages are immediately obvious: non-flammability and absolute environmental friendliness. These properties make it suitable for many applications where non-toxicity and hygiene are required. Examples include the food industry and the pharmaceutical and chemical industries.

However, water as a pressure medium is beset by certain problems due to its specific properties. Its viscosity is lower than that of mineral oil by one to two orders of magnitude. Its vapor pressure is far higher than that of highly viscous oil. Some undesirable phenomena occurring because of these characteristics are unfavorable tribological properties, corrosion, erosion through cavitation, and the incidence of microorganisms under conditions of inadequate maintenance.

This book addresses the particular characteristics of water hydraulics in a thorough manner. A sufficiently long service life of components presupposes that sophisticated corrosion resistant materials be used. Of particular importance for solving tribological problems is the application of polymeric coatings to sliding surfaces and seals. This allows axial piston displacement units to achieve very good volumetric and hydraulic-mechanical efficiencies. Besides pumps and motors, the book discusses valves, lines, containers, filters, and accumulators. In particular, the book contains many useful insights concerning the maintenance of water as a pressure medium and the monitoring of bacteriological contamination.

The author frequently refers to products of the Danfoss Company, within the framework of its Nessie® program, which has diligently and systematically brought water hydraulic components and systems to maturity. The primary objective of this development process was not so much the transmission of high power, but the utilization of water hydraulics' potential (in terms of power density, controllability, and transmittability of hydraulic power over medium distances) for applications requiring the environmentally friendly and safety properties of water as a pressure medium. For these applications, a pressure range of 140 to 160 bar is sufficient.

Summarizing, one can say that the result is a worthwhile book treating all important aspects of water hydraulics without troubling the reader with unnecessary details. Since many of the fundamental concepts are identical to those of oil hydraulics, the book may also serve as an introduction to the field of hydraulics as a whole. Both for students and professional engineers this book represents a suitable introduction to this particular field of drive technology. Specialists in the areas of food, pharmaceutical and chemical processing; water treatment; manufacturing; construction; mining and off-shore exploration; and transportation will also find much useful knowledge in this book.

Wolfgang Backé
em. Prof. Dr.-Ing. Dr. h.c. mult

PREFACE

A new epoch of pressure hydraulics, the revival of water hydraulics, has arrived. But it is much more than that, because water hydraulic control based upon newly developed standard components has evolved to meet the special requirements of many new and sensitive industrial applications. Several manufacturers of water hydraulic components and systems have their products on the market, and customers and users are curiously watching this emerging technology.

It is felt that because water hydraulic technology—at least in the near future—addresses mainly new application areas for hydraulic control automation rather than competing with the well-established oil hydraulic domain, there is a need for an introductory book on the basics of water hydraulic technology and its pros and cons.

The material in this book is organized and explained so as to be understood by an audience of various educational backgrounds, ranging from automation technicians, technical students, design engineers, and executives to researchers and professors.

The various topics are presented so that the differences between water hydraulics and oil hydraulics are clearly explained. Many design principles and solutions are similar or sometimes even identical in the two areas. Readers with experience in oil hydraulics may skip chapters 2 and 3, and sections 5.1–5.3, 6.1 and 9.1–9.2.

Many new ingenious water hydraulic products are surely going to be designed, developed and used in the coming years for a wide range of application areas. It is hoped that this book will prove its value to any technician/engineer who wants to understand, acquire and use water hydraulic systems.

In a new field such as water hydraulic control the information for a monograph like this has to be collected from several sources and people.

Every book has its genesis. This project was proposed to me by the President of the hydraulic division of Danfoss, Mr. Jørgen Mads Clausen, to whom I am most grateful. In the process of writing, the sales and marketing manager of the industrial hydraulics division at Danfoss, Mr. Jørgen W. Lorentzen, his colleagues, and several engineers from the water hydraulic components group offered me their valuable assistance, and they are all acknowledged for this support.

Finally, thanks to Mrs. Sofia Kroszynski, who helped with typing the text and preparing the drawings.

Erik Trostmann
Professor, Ph.D.

CONTENTS

1. INTRODUCTION

1.1 Power transmission

Power transmission systems are used to transfer power from an energy source, often called the prime mover, to a load in order to perform useful tasks or work.* Technically speaking, these tasks are handled by machines. The purpose of machines is to execute useful work processes throughout a certain, defined service life and with a certain, expected reliability.

From this point of view machines can be considered as more or less complex structures of energy sources, power transfer systems and load processes. In such structures power and information are transferred, modulated and transmitted. Any machine includes a feasible **power transmission system**.

The design of machines is dictated primarily by the functional processes to be carried out. Ideal design solutions aim at deriving a power transfer system that ensures an **optimal matching** of a selected (available) power source and the machine load process.

In Fig. 1.1 a schematic flow diagram of a general power transmission system is illustrated. The system or the drive transmits power from a prime mover, e.g., a diesel engine, to the load (i.e., the machine to be driven). The power generator generates power, which via a feasible modulation is transferred to the motor driving the load in the machine.

POWER TRANSMISSION SYSTEM (DRIVE)

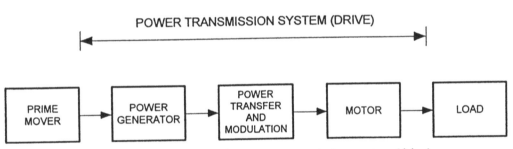

Fig. 1.1 Flow diagram of a power transmission system (drive)

Power transmission systems are often called drives and can be classified in three different categories depending on the input and output type of mechanical energy (ref. 17):

1. Rotational/rotational power, R ⇔ R (torque on rotating shaft/torque on rotating shaft)

2. Rotational/translation power, R ⇔ T (torque on rotating shaft/force on curvilinear motor)

3. Translational/translational power, T ⇔ T force on curvilinear motion/force on curvilinear motion)

The above categories of drives are shown in a schematic form in Fig. 1.2 as pure mechanical systems and hydraulic systems, respectively.

---

* For the definitions of *work* and *power* see section 1.3

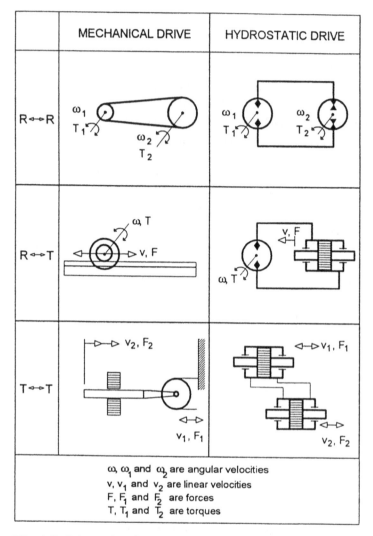

Fig. 1.2 Schematic of pure mechanical and hydrostatic drives

Often power transmission system are also classified in accordance to the physical principle applied in the drive:

M	-	Mechanical
H	-	Hydraulic
P	-	Pneumatic
EM	-	Electromechanical
E	-	Electronic

In practical applications the various physical principles are frequently mixed, such as in electrohydraulic systems.

When considering a certain application, it may be asked: Which one of the above-mentioned types of physical drives should be chosen as an optimal solution? A general unified answer to this question, however, cannot be given. Too many parameters influence the optimal choice of a drive to a specific application. But even so, it still makes sense to carry out comparisons between the primary properties of the different types of drives in order to identify the advantages and disadvantages of the individual type of drive.

In Fig. 1.3 a schematic comparison of important properties between the above-mentioned types of drives is given in a tabular form. It can be seen that hydraulic drives have several highly rated drive properties.

DRIVE PROPERTY	RATING OF PROPERTY		
	HIGH	MIDDLE	LOW
POWER TO WEIGHT	H	P	E,EM,M
TORQUE TO INERTIA	H	P	E,EM,M
SPEED OF RESPONSE	H	E,EM	P,M
CONTROLLA-BILITY	H,EM	E,EM,P	M
LOAD STIFFNESS	H	M	E,EM,P
VELOCITY RANGE	H,EM	E,EM	P,M

Fig. 1.3 Comparison of important drive properties for various types of systems

In the following paragraphs it is briefly explained why hydraulic drives in many cases are superior to other types of drives, especially electrical drives.

1. The minimum size of an electrical motor is usually determined by the maximum magnetic flux density required and the maximum heat dissipation allowable in the motor insulation material. In comparison the minimum size of a hydraulic motor is limited only by structural load considerations for the motor material. This leads to power/weight and torque/inertia ratios for hydraulic motors which at higher power ranges are orders of magnitude higher than for electric motors. In Fig. 1.4 the compactness of a hydraulic motor compared with an electric motor and a diesel engine is shown. The three units are of the same power rating.

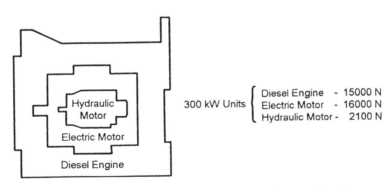

300 kW Units { Diesel Engine - 15000 N
Electric Motor - 16000 N
Hydraulic Motor - 2100 N

Fig. 1.4 Comparison of the size of a diesel engine, an electric motor and a hydraulic motor (ref. 9)

2. Secondly, a hydraulic motor is mechanically stiff in its load-bearing capability, in comparison with an electric motor. This means that the hydraulic medium can lock the motor in position when the motor is stopped. In comparison, the magnetic field in the electric motor is soft and compliant.

3. Thirdly, the hydraulic drive offers a high attainable speed of response, in other words, a high acceleration. Especially when large power requirements (>10 kW) are needed and a high speed of response is specified (>100 Hz), the hydraulic drive may be the only available solution.

In many respects hydraulic (and pneumatic) power transmission systems are placed somewhere between electrical and mechanical systems. It is easier to transmit hydraulic power over appreciable distances than mechanical power but it is easiest to transmit electrical power.

A leading advantage of a hydraulic power drive is its high versatility for a functional integration with automatic machinery due to the high ratio of torque to inertia and power intensity to weight.

Hydraulic drives also possess some disadvantages. The most serious of these are the messy leakage into the environment from hydraulic systems using oil as pressure medium and the serious danger of fire and explosion when using oil hydraulic power.

The recent development and marketing of Nessie® water power technology has completely removed these risks by using **pure tap water** as the hydraulic pressure medium. This book aims to promote "best practice" approach to water technology to make water hydraulic design an advanced engineering discipline.

The issue of safety in the design and use of power transmission systems is, as in any other engineering domain, of utmost importance. The revolutionary arrival of the Nessie® technology has improved safety without sacrificing excellence in design, workmanship, or maintenance.

1.2 History of hydraulics

Water has been used as a means of transmitting power from very early times. For example, **Ctesibus**, the son of a barber in Alexandria, has been accredited with inventing the first pump in the second century B.C. This device was a two-piston, positive-displacement unit used for pumping water.

Shortly after the development of the first steam engines in the 18th century a considerable interest for transmission of power using a **liquid under pressure** arose, because this technique was relatively accurate and economic. In 1795 **Joseph Bramah** was granted a British patent (Brit. pat. 2045) for his invention of the first water hydraulic press (ref. 13). The industrial revolution in the 1850's in the United Kingdom led to the use of fluid power for powering presses, elevators, cranes, extruding machines etc. In London and other cities as well, central industrial hydraulic networks were established. The networks distributed several thousands of horsepower. The network **London Hydraulic Power Co.** is still in service.

Around the year of 1900, however, a very strong development of electrical power supply networks took place, featuring the power supply and control of electric motors. This development retarded and in some cases even reversed the spread of hydraulic power and control for several decades.

In 1906 the first oil hydraulic system was introduced. A hydrostatic transmission with axial machines was developed by **Janney** using oil as the fluid medium. The transmission replaced the electromechanical system for elevating and controlling warship guns. Radial piston machines for oil hydraulics were introduced in 1910 by **H. S. Hele-Shaw** and in 1922 by **Hans Thoma**. The latter also developed the first variable displacement pump with a tilting head axial piston type. In 1930 **Harry Vickers** invented the pilot-operated pressure relief valve.

However it was not until the years after the Second World War that the use of hydraulic power transmission and control was reborn and really expanded in the control of machines. New hydraulic components and systems were invented and used in numerous applications. An important breakthrough was the development of solenoid-operated hydraulic directional valves for the **automatic** control of machines.

Some of the important developments during the 1950's were swash plate hydraulic machines and **electrohydraulic servovalves** for servo systems.

In 1978 the Royal Navy in the United Kingdom contracted the National Engineering Laboratory (NEL) to develop an undersea tool using seawater as a hydrostatic medium for power transmission. Later Shell/Esso contracted NEL to continue the development of undersea tools and a power pack based on seawater. In 1987 Scot-Tech, a subsidiary of the UK firm Fenner, was set up to continue the development of water fluid power components (ref. 4).

In 1994 the Danish company Danfoss, for many years a well-known oil hydraulic component manufacturer, introduced a complete new water hydraulic system technology, Nessie®, using **pure tap water** as the fluid medium without sacrificing the unique hydraulic control performance characteristics of oil hydraulic systems.

The historical highlights in the use of hydrostatic power are illustrated in Fig. 1.5. In early days water-based hydraulics was universal. But since the beginning of this century oil hydraulic systems have become more and more dominant. Recently, however, it has become clear that the new water hydraulic system technologies, like Nessie®, may revolutionize hydraulic power transmission systems technology in the years to come.

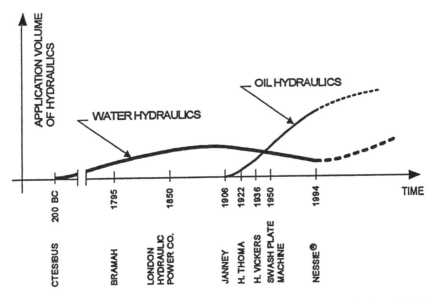

Fig. 1.5 Historical development of hydrostatic power transmission (ref. 3)

In short the Nessie® water hydraulics technology, by using tap water as the pressure medium, has overcome the two, increasingly important, very risky disadvantages of traditional oil hydraulics: pollution caused by the leakage of oil into the environment and the risk of fire and explosion from using mineral oil.

At the same time the Nessie water hydraulics technology, by using ingenious design of bearings in pumps and motors and by selection of nontraditional materials for the components, has overcome the major drawbacks of the old water hydraulics: the problem of having correct lubrication of the moving parts in the components and the problem of avoiding corrosion.

The advent of Nessie water hydraulics may open up for new hydraulics applications, e.g., in food production, medicine manufacturing, etc., where oil hydraulics by and large has hitherto been avoided. This is indicated in Fig. 1.5 by the increase in the curve for water hydraulics after 1995.

In the majority of hydraulic power transmission systems (hydraulic control systems) the primary type of energy contained in the system is pressure energy. Such systems are called hydrostatic systems, as distinguished from hydrokinetic systems. In the latter the major energy component is kinetic energy in the flowing fluid. In hydrostatic systems the motors and pumps used are based upon the displacement principle, i.e., they are positive-displacement machines.

Generally speaking, a hydraulic control system is characterized by four basic types of components, as shown in Fig. 1.6. These components are hydraulic power generators (pumps), hydraulic control components or modulators (valves), hydraulic transmission lines (tubes, hoses, fittings, etc.) and hydraulic actuators (single- or double-acting cylinders and rotary motors).

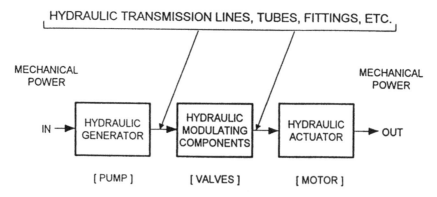

Fig. 1.6 Hydraulic power transmission system

The power transmitted in a hydraulic control system is computed as the **product of the volume flow rate and the pressure**. For the simple hydraulic system shown in Fig. 1.7 the power transmitted in the system is shown in Fig. 1.8 as a function of the position in hydraulic path through the system. The individual sections are described with the same position numbers in Fig. 1.7 and Fig. 1.8. For the system shown the volume flow rate Q is constant and the pressure differences, Δp_1 and Δp_2, are proportional to hydraulic input power and output power, respectively.

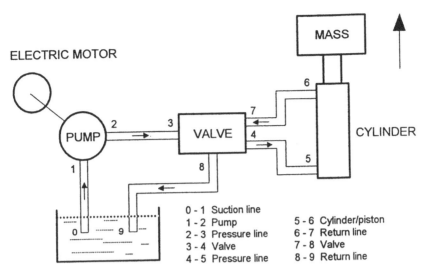

ELECTRIC MOTOR

MASS

PUMP   2   3   VALVE   6   7   4   CYLINDER   5   8   1

0 - 1 Suction line	
1 - 2 Pump	5 - 6 Cylinder/piston
2 - 3 Pressure line	6 - 7 Return line
3 - 4 Valve	7 - 8 Valve
4 - 5 Pressure line	8 - 9 Return line

Fig. 1.7 Simple hydraulic system

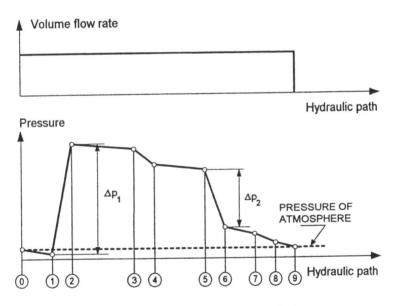

Volume flow rate

Hydraulic path

Pressure

Δp_1

Δp_2

PRESSURE OF
ATMOSPHERE

Hydraulic path

⓪   ①②   ③④   ⑤⑥   ⑦⑧⑨

Fig. 1.8 Hydraulic power transmission

Each of the different physical power systems has its strengths and weaknesses. While a particular system may be dominant for a certain application, there exists a dynamic balance due to the continuous appearance of new applications and new products. Although it can be useful to compare the relative balance among these systems at a given time, it should be noted that for many applications the different power transmissions are not compatible. Therefore the graphs in Fig. 1.9 must be taken as a general guide, and specific applications should be compared on a case-by-case basis.

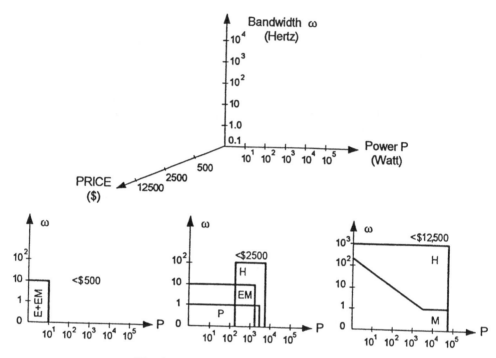

Fig. 1.9 Comparison of different drives

In the graphs in Fig. 1.9 the three following basic parameters are used as main properties to describe a drive: (1) power in watts (P), (2) bandwidth in Hertz (ω), and (3) price in dollars ($\$$). These are used as coordinates in a 3-dimensional Cartesian coordinate system for describing different types of drives. Note the logarithmic scale for the bandwidth and the power.

Fig. 1.9 shows three cross-sections corresponding to a system price in dollars, respectively $1000, $3000, and $12,500; the corresponding (P,ω) plane is illustrated at the bottom of the figure. For each of the three (P,ω) planes the significant application areas of the different drives are indicated.

In spite of the general caution noted on p. 7, the areas where hydraulic systems are dominant are clearly identifiable.

It is probably neither possible nor necessary to identify all the applications where hydraulics may be applied. The potential of applications using hydraulic power transmission is still expanding. However, a certain terminology for categorizing application areas for hydraulic control is often used in practice, and therefore it is appropriate to mention it here:

(a) Hydraulic control systems and equipment used in machine tools, presses, plastic injection machines and extruders, robots and other production equipment are termed **industrial hydraulics**.

(b) Hydraulic systems and equipment used in construction machinery, agricultural machines, automobiles and trucks are termed **mobile hydraulics**.

(c) Hydraulic systems for the control and maneuvering of airplanes and aerospace vehicles are often denoted as **aircraft hydraulics**.

(d) Hydraulic systems applied on-board ships for driving winches, cranes, hatches, and for maneuvering of valves, etc., are termed **ship hydraulics**.

(e) Hydraulic systems used on off-shore platforms, for example, for oil drilling, are termed **off-shore hydraulics**.

1.3 Elementary hydraulic principles

A fluid is infinitely flexible and, like a solid, to a great extent incompressible. It can change its shape in accordance with the containment. It can be divided into two or more flow ways and carry out work at different places. Further, it can change flow rate in a flow way depending on the cross-sectional area of the flow way, and it can exert a force in any direction. No other medium exhibits the same positive integrity, accuracy and flexibility in controlling power with minimal weight and volume.

In hydraulics the two most important physical variables are pressure p and flow rate Q. Considering power transmission systems the variables (p, Q) are analogous to the mechanical variables force F and velocity v or torque T and speed of rotation n.

In the following sections characteristics of pressure and flowrate will be considered in detail.

PRESSURE

The static pressure in a liquid is governed by Pascal's law, which may be described by the following three statements:

(a) Ignoring the weight of the fluid and assuming that the fluid is at rest, the static pressure in the fluid will be the same everywhere in the fluid (see Fig. 1.10).

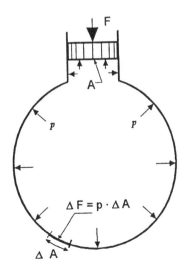

Fig. 1.10 Pascal's law

(b)   This static pressure is transmitted instantaneously and remains undiminished in all directions.

(c)   This static pressure will always be perpendicular to any bounding surface.

The static pressure p is defined as force per area (see Fig. 1.10):

$$p = \frac{F}{A}$$

The force ΔF at a surface element ΔA in Fig. 1.10 is perpendicular to the surface element and is computed by:

$$\Delta F = p \cdot \Delta A$$

The total force is the sum of force elements given by:

$$F = p \cdot \Sigma \Delta A = p \cdot \int dA = p \cdot A$$

For a given force direction the area A equals the projection of the containment surface in this direction, even when the surface is curved.

Pressure and force are measures of effort. **Work** W carried out is equal to the force multiplied by the distance moved s, i.e.:

$$W = F \cdot s$$

The concept of work is, by definition, independent of time. Work per time is defined as power P.

Since a fluid in practical terms can be considered incompressible, mechanical forces may be transmitted, amplified and controlled (see Fig. 1.11). The figure shows the simplest form of a hydrostatic force multiplier (amplifier). Two cylinders X and Y contain pistons that fit watertight into the cylinders but that are still able to move without friction. The cylinders are connected at the closed ends via a pipe connection. The cylinders are filled with a liquid (e.g., water) and dimensioned to allow for some vertical movement.

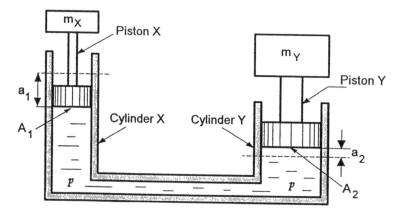

Fig. 1.11 Simple force multiplier

From Pascal's law it is seen that pressure between the cylinders can be transmitted instantaneously and unchanged. The piston forces F_X and F_Y for cylinders X and Y, respectively, can be computed by using Newton's second law, as follows:

$$F_X = m_X \cdot g \qquad \text{and} \qquad F_Y = m_Y \cdot g$$

where m_X and m_Y are the respective masses of the pistons X and Y, and g is the acceleration of gravity.

Assuming that a piston moves slowly and with no friction, the pressure p is computed as follows:

$$p = \frac{F_X}{A_1} = \frac{F_Y}{A_2}$$

which gives:

$$F_X = \frac{A_1}{A_2} F_Y$$

where $\frac{A_1}{A_2}$ is considered as the **force amplification.**

Assuming further that the liquid is incompressible and that the cylinder as well as the connections are of non-elastic material, a movement a_2 of the piston Y downward (see Fig. 1.11) will create a movement a_1 of piston X upward given by:

$$a_1 = \frac{A_2}{A_1} a_2$$

Further the work W_X and W_Y carried out by pistons Y and X, respectively, will be:

$$W_X = F_X \cdot a_1 \ (=) \ W_Y = F_Y \cdot a_2$$

and if the movement is done over a period of time equal to t, the **power** E related to the movements of the positions equals:

$$E = \frac{F_X \cdot a_1}{t} = \left(\frac{F_y \cdot a_2}{t} \right)$$

or

$$E = \frac{p \cdot A_1 \cdot a_1}{t} = p \cdot Q$$

where the liquid volume flow rate (volume per time) from cylinder Y to cylinder X is Q, given by:

$$Q = \frac{A_1 \cdot a_1}{t}$$

Note that the **hydraulic power** transmitted from, say, cylinder X to cylinder Y is computed as **pressure p times flow rate Q.**

FLOW

Considering a flow channel (a pipe) with variable cross-sectional area and assuming an ideal fluid, the volume flow rate Q in each cross-section of the pipe will be the same in accordance with the **continuity equation**:

$$Q = v_1 A_1 = v_2 A_2 = \dots v_n A_n$$

where $v_1, v_2, \dots v_n$ and $A_1, A_2, \dots A_n$ are the fluid velocities and the respective cross-sectional areas for cross-sections $1, 2, \dots n$.

The fluid velocity is inversely proportional to the cross-sectional area:

$$\frac{v_1}{v_2} = \frac{A_2}{A_1}$$

When the fluid is ideal, i.e., the weight density γ is constant and the flow is frictionless, the same mass of the fluid per time flows through each cross-section of the pipe. With reference to Fig. 1.12 and in accordance with the **Bernoulli equation,** the mass unit in each cross-section contains the same amount of energy. The equation can be expressed by:

$$\frac{v_1^2}{2g} + \frac{p_1}{g} + h_1 = \frac{v_2^2}{2g} + \frac{p_2}{g} + h_2 = constant$$

where the indexes (1) and (2) refer to cross-sections (1) and (2), γ equals the specific weight of the fluid, v_1 and v_2 are the average velocities, p_1 and p_2 the pressures, and h_1 and h_2 the elevations.

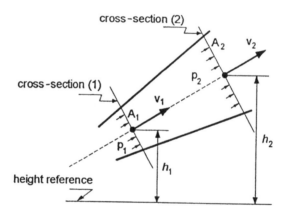

Fig. 1.12 Bernoulli flow

In general the Bernoulli equation can written by:

$$h = h_{dyn} + h_{stat} + h_{pot} = constant$$

where h is the total energy content and

$$h_{dyn} = \frac{v^2}{2g} = \text{kinetic energy}$$

$$h_{stat} = \frac{p}{\gamma} = \text{static pressure energy}$$

$$h_{pot} = \text{potential energy}$$

The energy terms are expressed by pressure height (a length).

In a real fluid flow where friction in the fluid itself and along the pipe wall is present, a certain energy loss, h_v, will occur when fluid flows from cross-section A_1 to A_2. Furthermore, it can be assumed that the potential energy h_{pot} is unchanged, i.e., $h_1 = h_2$, because the level of the cross-sections is constant. Now, a modified Bernoulli equation can be expressed as:

$$\frac{v_1^2}{2g} + \frac{p_1}{g} = \frac{v_2^2}{2g} + \frac{p_2}{g} + h_v$$

The energy loss h_v is converted into heat in the fluid, and a corresponding pressure drop occurs. Assuming the flow velocity is constant, i.e., $v_1 = v_2 = v$, and assuming that the pipe is circular and has an internal, smooth surface, the pressure loss Δp between cross-sections (1) and (2) in Fig. 1.12 can be empirically expressed by:

$$\Delta p = p_1 - p_2 = \frac{\lambda \cdot \gamma \cdot v^2}{2g} \cdot \frac{L}{D} + \frac{\psi \cdot \gamma \cdot v^2}{2g}$$

where λ = dimensionless friction factor for straight pipes
        γ = specific weight
        ψ = dimensionless resistance coefficient
        L = length of pipe
        D = internal diameter of pipe
        v = average velocity

In the above equation the first term on the right-hand side expresses the energy loss in the flow due to friction, and the second term expresses energy loss in the flow caused by bends, fittings and sudden changes in flow cross-section. These losses are often termed minor losses.

The friction factor λ depends on the Reynolds number R (dimensionless) for the actual fluid flow, whether it is laminar or turbulent flow. This relationship is described in the following paragraphs.

The Reynolds number for a pipe with circular cross-section is defined by:

$$R = \frac{v \cdot D}{\nu}$$

where

        v = average velocity
        D = internal diameter of pipe
        ν = kinematic viscosity of fluid

The flow in a pipe may be either laminar or turbulent depending on whether the Reynolds number is smaller or bigger than a critical value R_{cri}. For $R < R_{cri}$ the flow is laminar and for $R > R_{cri}$ the flow is turbulent. For circular and smooth pipes R_{cri} equals 2000–2300. For non-circular cross-sections of a pipe other values for R_{cri} must be used. Note that in **water hydraulic systems flow conditions for turbulent flow will be predominant**.

From empirically derived relations the friction factor λ can be computed from the following expressions:

$$\lambda = \frac{64}{R} \qquad R < R_{cri}$$

$$\lambda = 0.3164 \cdot R^{-0.25} \qquad R > R_{cri} \qquad \text{(Blasius' formula)}$$

The above expressions are shown in graphical form in Fig. 1.13. When full laminar flow has not been fully developed and at high values for the Reynolds number, the values for λ are slightly modified, as shown by the dashed lines in Fig. 1.13.

Friction factor λ

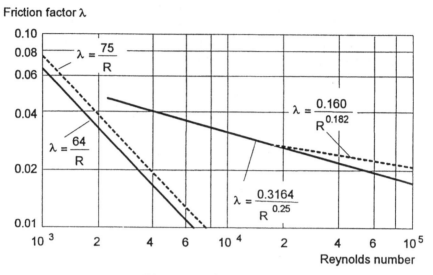

Fig. 1.13 Friction factor λ

The resistance coefficient ψ is also determined by empirical means and may be transformed to equivalent pipe length. Quite often the suppliers for bends, fittings, etc., offer experimental values for pressure loss in such typical components, as shown in Figs. 1.14 and 1.15. Resistance coefficients due to abrupt changes in pipe cross-sections and in geometry at pipe entrances can also be computed by empirical formulas.

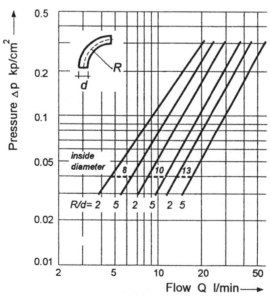

Fig. 1.14 Pressure drop for pipe bends

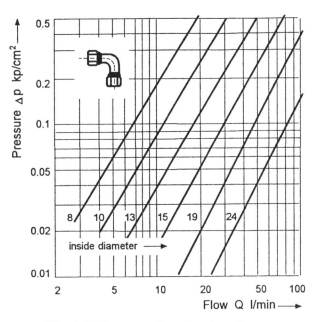

Fig. 1.15 Pressure drop for pipe elbow

Assuming incompressible **laminar** flow with friction through a circular smooth pipe, the volume flow rate can be computed using the well-known **Hagen-Poiseuille formula**:

$$Q = \frac{1}{128} \cdot \frac{\pi \cdot D^4 (p_1 - p_2)}{\mu L}$$

where

Q = volume flow rate through pipe
D = pipe diameter
$p_1 - p_2$ = pressure drop along the pipe length
L = pipe length
μ = dynamic viscosity

For flow through channels of non-circular cross-sections formulas equivalent to the one above exist and can be found in ref. 10.

Assuming incompressible **turbulent** flow with friction through a circular smooth pipe the volume flow rate can be expressed by using **Blasius' formula** for the friction factor λ (ref. 10):

$$Q^{1.75} = \frac{D^{4.75}}{0.242 \cdot \mu^{0.25} \cdot \rho^{0.75}} \cdot \frac{p_1 - p_2}{L}$$

where

Q = volume flow rate through pipe
D = pipe diameter
$p_1 - p_2$ = pressure drop along pipe length
L = pipe length
μ = dynamic viscosity

Orifice flow is used very often to control fluid power, for instance, in many valve constructions. An orifice is a sudden restriction of short length introduced into a flow passage (see Fig. 1.16). It may be of constant or variable cross-sectional area. Assuming an ideal fluid, the Bernoulli equation leads to the following formula for volume flow rate through an orifice:

$$Q = C_d \cdot A_0 \cdot \sqrt{\frac{2}{\rho}(p_1 - p_2)}$$

where Q = volume flow rate through orifice
A_0 = cross-sectional area of orifice   ($A_0 << A_1$)
ρ = specific mass of the fluid
$p_1 - p_2$ = pressure drop across the orifice
C_d = discharge coefficient (empirical)

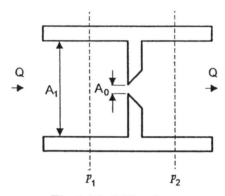

Fig. 1.16 Orifice flow

Most often the orifice area A_0 is small compared with the cross-sectional area A of the passage (see Fig. 1.16). When this is the case, the value of C_d is approximately 0.60. In other cases dedicated experiments are needed to define the correct value of the discharge coefficient.

Example 1.1

With reference to Fig. 1.11 the following quantities are assumed:

$$\begin{aligned}
F_X &= 500 \text{ N} \\
A_1 &= 5 \text{ cm}^2 \\
A_2 &= 20 \text{ cm}^2 \qquad t = 0.01 \text{ sec} \\
s_1 &= 0.5 \text{ cm}
\end{aligned}$$

Then the following values can be computed:

$$F_Y = \frac{A_2}{A_1} \cdot F_X = \frac{20}{5} \cdot 500 = 2000 \text{ N}$$

$$p = \frac{F_X}{A_1} = \frac{500}{0.0005} \cdot 10^{-5} \text{ bar} = 10 \text{ bar}$$

$$a_2 = \frac{A_1}{A_2} \cdot s_1 = \frac{5}{20} \cdot 0.5 = 0.125 \text{ cm}$$

$$W_x = F_X \cdot s_1 = 500 \cdot 0.005 = 2.5 \text{ J} \ (= W_Y)$$

$$Q = \frac{A_1 \cdot s_1}{t} = \frac{5 \cdot 0.5}{0.01 \cdot 1000} = 0.25 \frac{1}{\text{sec}}$$

$$E = \frac{p \cdot A_1 \cdot s_1}{t} = \frac{10 \cdot 10^5 \cdot 0.0005 \cdot 0.005}{0.01} = 250 \text{ watt}$$

or

$$E = p \cdot Q = 10 \cdot 10^5 \cdot 0.25 \cdot 10^{-3} = 250 \text{ watt}$$

Example 1.2

A hydraulic fluid is pumped through a 300-m-long, horizontal, circular pipe with a diameter of 5 mm. The flow is $Q = 1.23$ cm³/sec, the total pressure drop along the pipe is $\Delta p = 19.62$ N/cm². Determine the kinematic viscosity v in cS, when the mass density is 860 kg/m³.

The condition for laminar flow in the pipe requires

$$R = \frac{\rho \cdot v \cdot D}{\mu} < 2000$$

Therefore, assuming

$$\mu > \frac{\rho \cdot v \cdot D}{2000} = \frac{860 \cdot \dfrac{123 \cdot 10^{-6} \cdot 4}{\pi \cdot 0.005^2}}{2000} = 0.1347 \cdot 10^{-3} \frac{\text{N sec}}{\text{m}^2}$$

The Hagen-Poiseuille formula gives

$$\mu = \frac{1}{128} \cdot \frac{\pi \cdot D^4 \cdot \Delta p}{Q \cdot L}$$

$$= \frac{\pi \cdot 0.005^4 \cdot 19.62 \cdot 10^4}{128 \cdot 1.23 \cdot 10^{-6} \cdot 300} = 0.00816 \frac{\text{N sec}}{\text{m}^2}$$

$$v = \frac{\mu}{\rho} = \frac{0.00816}{860} \frac{\text{m}^2}{\text{sec}} = 9.5 \text{ cS}$$

Example 1.3

The flow through an orifice is given by the following parameters:

$$C_d = 0.60 \ (\text{empirical value})$$
$$A_o = 0.5 \text{ mm}^2$$
$$\rho = 1000 \text{ kg/m}^3$$
$$p_1 - p_2 = 100 \text{ bar}$$

The flow rate is computed by:

$$Q = C_d \cdot A_o \sqrt{\frac{2}{\rho}(p_1 - p_2)} = 0.60 \cdot 0.5 \cdot 10^{-6} \cdot \sqrt{\frac{2}{1000} \cdot 100 \cdot 10^5} = 42.4 \cdot 10^{-6} \frac{\text{m}^3}{\text{sec}} = 42.4 \frac{\text{cm}^3}{\text{sec}}$$

2. STANDARD SYMBOLS FOR FLUID POWER SYSTEMS

Through the 1930's and 1940's more and more hydraulic "standard" components, such as valves, pumps and cylinders, became available on the market. Many users felt an increasing need for "neutral" symbols for defining hydraulic components when hydraulic diagrams were worked out during the engineering design and acquisition process. A neutral symbol means a universally accepted and vendor-independent symbol.

The discussion on fluid power symbols and standards was initiated in the United States by the establishment of the **Joint Industry Conference**, JIC, in 1944. Representatives of the automotive industry, the fluid power component manufacturers and the machine tool builders formed the JIC. The first JIC-defined hydraulic symbols and standards were released in 1948. Revisions were adapted and released during the following years, until the last JIC standard was issued in 1959. Then the national standardization organisations took over. In 1958 the American Standards Association, ASA, adapted officially a slightly revised version of the JIC standards for fluid power diagrams. In 1965 ASA approved an American standard for hydraulics.

A European organization for hydraulics and pneumatics was formed in 1962 with the name **Comité Européen des Transmissions Oléohydrauliques et Pneumatiques,** CETOP. Leading European manufacturers of hydraulic and pneumatic equipment are members of CETOP, which has worked out and published a series of recommendations in the fluid power area. Many of these are approved by the **International Organization for Standardization**, ISO, and national standardization organizations as well.

Hydraulic systems and circuits are normally described in diagrams where graphical standard symbols are used in accordance with those of the ISO standard 1219. These symbols are normally referred to as ISO/CETOP symbols. A short introduction and explanation to these will be given in the following paragraphs. In Appendix D a list of the most commonly used symbols is given.

The graphic symbols consist of one or more basic symbols and of supplementary symbols. The basic symbols illustrate the basic function of the component, and the supplementary symbol illustrates the principle of operation.

The symbols can be drawn at any scale or orientation suitable for clear interpretation and by using a convenient line width.

The symbols illustrate hydraulic connections, flow paths and functions of components. But they do not show the design or the construction principle. The symbols do not indicate physical dimensions, flow and pressure ratings or actual location of ports and control elements on the actual component. The graphical symbols in a circuit are drawn to show normal or neutral state condition of the component except when the hydraulic circuit diagram is used to illustrate various phases of the circuit operation.

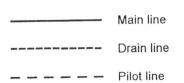

Fig. 2.1 Signatures for flow line types

In a hydraulic diagram a hydraulic **main flow connection** is indicated by a full line (see Fig. 2.1). The connection may be a suction, pressure or return line.

A **drain line** is illustrated by a short-dashed line and carries leakage flow from a component, for example, a hydraulic motor, back to the reservoir.

A **pilot line** is illustrated by a long-dashed line. It carries only minimal flow and is used to transmit a pressure signal from one location to another.

Directional control valves are used to control which main lines that may carry pressure flow and return flow. The basic symbol for a directional control valve consists of a number of rectangles corresponding to the number of states or functional positions of the valve (see Fig. 2.2).

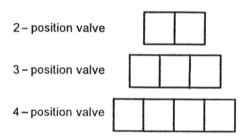

2 – position valve

3 – position valve

4 – position valve

Fig. 2.2 Basic symbols for directional control valves

The principle of operation is illustrated in Fig. 2.3, where the internal connections in a 3-position and a 2-position valve are illustrated. Note that the external connections to the hydraulic circuit are indicated on only one of the rectangles (states). The internal flow ways in the valve are indicated by arrows. The external connections are labelled A and B for cylinder (motor) connections, P for pressure line connection and R for return line connection. Note also that many different combinations of internal flow ways are possible (see Appendix D).

Sometimes directional control valves are also designated with the fundamental number of external connections as well as the number of position states. For example, the valves shown in Fig. 2.3 have four and three external connections, respectively, and therefore can be called 4-port, 3-position and 3-port, 2-position directional control valves.

The symbol at the end of the return line connection R indicates that this line is connected to the hydraulic reservoir.

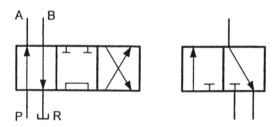

Fig. 2.3 Symbols for (left) a 3-position and (right) a 2-position directional control valve

To illustrate the content of the symbol in Fig. 2.3, a real spool-type directional control valve is shown in Fig. 2.4. The graphical symbol in Fig. 2.3, however, is generic and may illustrate other designs as well.

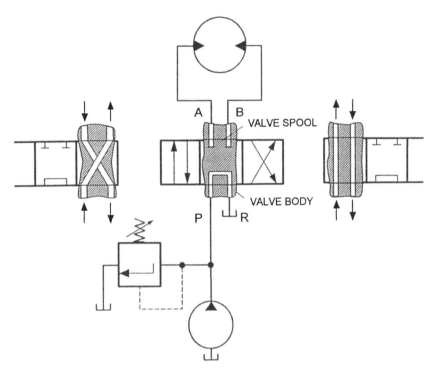

Fig. 2.4 The physical channels in a 3-position, 4-port directional valve

Directional control valves may be operated by different methods: manually, mechanically, electrically, hydraulically and pneumatically. These methods are illustrated in the graphical symbols shown in Fig. 2.5.

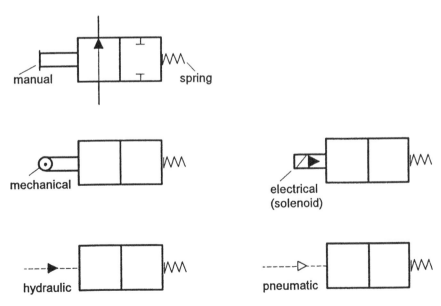

Fig. 2.5 Symbols for various operational principles for directional control valves

Pressure-control valves can be continuously positioned between a fully open and a fully closed state. The graphical symbol (see Fig. 2.6) consists of one rectangle and one internal flow way.

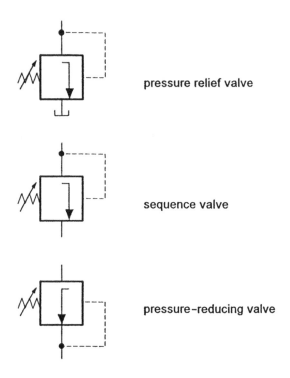

pressure relief valve

sequence valve

pressure–reducing valve

Fig. 2.6 Symbols for pressure-control valves

In the pressure-relief valve symbol the direction of the flow path arrow is kept away from the outlet port by a spring. When the inlet pressure exceeds the preset spring load, the pilot connection of the valve opens for flow to the outlet port. The spring is pre-loaded, corresponding to the desired relief pressure. The arrow at the spring indicates the pre-setting of the spring load. Note also that the connection of two hydraulic flow ways is indicated by a dot where the line meets.

The function of a sequence valve is similar to a pressure-relief valve, except that another component is connected to the outlet port.

The pressure-reducing valve is also in principle functionally similar to a pressure-relief valve. The pressure-reducing valve maintains the reduced pressure at the outlet port at a constant pressure (represented by the spring load) and independent of flow rate through the valve. The reduced pressure will, of course, always be lower than the input pressure to the valve.

In Fig. 2.7 a **check valve** or a non-return valve is shown. Symbolically the valve consists of a spring-loaded ball that, depending on the direction and magnitude of the pressure drop, either is closed against a seat or is open and allows for a flow in the direction of the arrow. In practice the check valve can be a ball valve, a poppet valve, etc.

The check valve may be of a type that is remotely operated by a pilot pressure. The pilot pressure can either block or open for flow in either direction.

direction of flow

Fig. 2.7 Symbol for a check valve

In principle a **flow control valve** introduces a restriction in the flow line by which the flow rate is controlled. The restriction can be a fixed or an adjustable orifice, shown symbolically in Fig. 2.8. Sometimes the orifice is combined with a check valve whereby flow is allowed in only one direction. When such combinations of more hydraulic functions in practice are carried out in one physical unit they should be indicated symbolically by an envelope signature as a dashed-dotted line (see Fig. 2.8).

More advanced flow control valves have provisions for pressure compensations and viscosity compensations such that the flow rate through the valve is held constant and independent of variations of the pressure drop across the valve and/or temperature variations of the pressure medium. The symbol for an adjustable flow control valve with pressure and viscosity compensations is shown in Fig. 2.8.

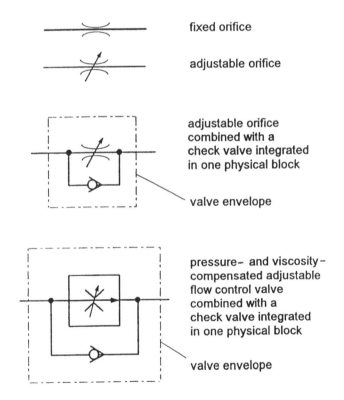

fixed orifice

adjustable orifice

adjustable orifice
combined with a
check valve integrated
in one physical block

valve envelope

pressure– and viscosity–
compensated adjustable
flow control valve
combined with a
check valve integrated
in one physical block

valve envelope

Fig. 2.8 Symbols for various flow control valves

The basic graphical symbol for **pumps** and rotary **motors** is a circle. Small triangles (see Fig. 2.9) indicate the direction of fluid flow through these components

In Fig. 2.10 symbols for hydraulic **cylinders** (linear motors) are shown. The cylinder block, the piston and the piston rod are indicated schematically. Examples of double-acting cylinders with single-end rod, with and without end cushioning, are shown.

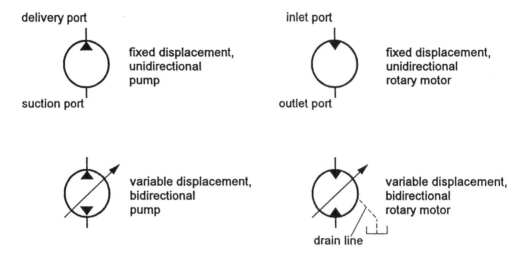

Fig. 2.9 Symbols for pumps and motors

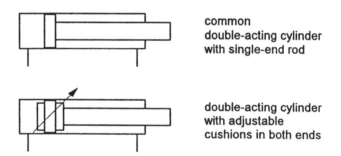

Fig. 2.10 Symbols for double-acting cylinders

A **simple example** of a hydraulic circuit graphically illustrated by the above standard symbols is shown in Fig. 2.11. The diagram indicates the cylinder control circuit given in Fig. 1.7. The circuit provides for extension and retraction of the piston in a double-acting cylinder controlled by a three-position direction control valve manually operated.

Whenever possible, the standard symbols (see Appendix D) are used in diagrams for hydraulic circuits in this book. From such diagrams the functions of the hydraulic systems can be derived and understood. When such diagrams become complex, where the hydraulic system contains several cylinders and/or motors controlled by an involved valve circuitry, state diagrams and timing diagrams are used to describe the functional sequence of the system function. A treatment of this latter technique is beyond the scope of this book (see ref. 6).

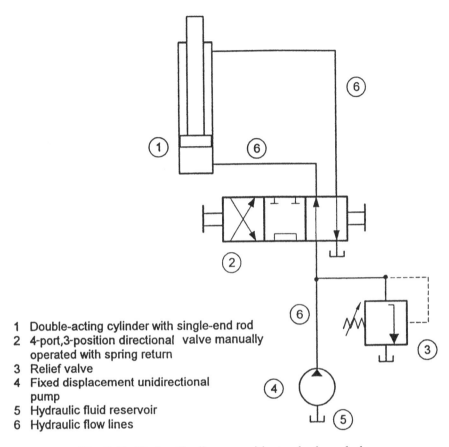

1 Double-acting cylinder with single-end rod
2 4-port,3-position directional valve manually
   operated with spring return
3 Relief valve
4 Fixed displacement unidirectional
   pump
5 Hydraulic fluid reservoir
6 Hydraulic flow lines

Fig. 2.11 Hydraulic diagram with standard symbols

3. SIMPLE HYDRAULIC CIRCUITS DESCRIBED BY STANDARD SYMBOLS

Hydraulic drives can generate almost any type of linear or rotary motion using a set of commercial standard components. In principle any hydraulic system is configured as shown in Figs. 1.6 and 1.7. The pump generates a flow of hydraulic fluid, which is pumped via pipe connections through suitable directional, pressure and/or flow control valves to a motor/cylinder and then back to the reservoir. Note that the pump generates a fluid flow, and the fluid pressure is generated by the resistance that the flow meets when it flows through the hydraulic system. The pressurized fluid flow per time delivers fluid power that is transformed to useful work output from the hydraulic motor/cylinder executing torques/forces and velocities upon load motions.

In this chapter a number of simple illustrative hydraulic circuits are described in order to introduce the functions of common commercial hydraulic standard components and familiarize the reader with the use of standard hydraulic symbols and signatures for the description of hydraulic systems on graphical diagrams.

Note that flow, pressure, speed and torque/force ratings of the components as well as their physical size and material or type of hydraulic fluid are not considered in this chapter.

A basic hydraulic system

A basic hydraulic system is shown in Fig. 3.1 in order to illustrate how hydraulic activation operates. A pump with fixed capacity pumps fluid from a reservoir into a single-acting, single-rod hydraulic cylinder. The piston moves upward as long as the pump is working and the valve is closed. The movement stops when the pump is stopped.

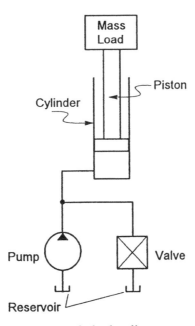

Fig. 3.1 Basic hydraulic system

When the pump is stopped, the piston can move downward if the valve is opened, thereby allowing the fluid from the cylinder to return to the reservoir. This assumes that the load on the piston is large enough to overcome friction forces acting upon the piston.

Basic hydraulic control

In order to take full advantage of the capabilities of the hydraulic power technology, a more advanced system is required (Fig. 3.2). The system contains several control features that make hydraulic control much more versatile, including a double-acting hydraulic cylinder and a manually operated, spring-centered, 4-port, 3-position directional control valve. The motion of the cylinder piston can be stopped in mid-stroke if the center position of the valve selected has the two cylinder ports blocked, i.e., the valve is of the closed-center type. Furthermore, in this valve position, the hydraulic fluid from the pump circulates at low pressure back to the reservoir, making it possible for the pump to run continuously.

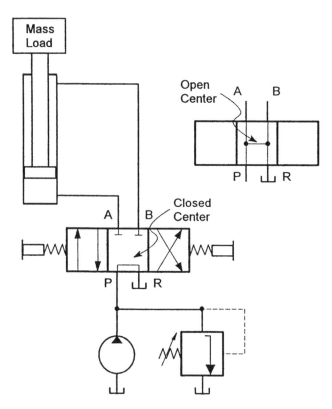

Fig. 3.2 Use of some basic hydraulic control capabilities

When the valve in Fig. 3.2 is in its right position, the piston moves downwards, and when it is in its left position, the piston moves upwards.

If the piston reaches one of the ends of the cylinder or the work load stops it, pressure liquid from the pump flows through the relief valve back to the reservoir. The pressure equal to the relief valve setting (adjustable) is then maintained on the piston by the pump. The generated fluid power is lost in the relief valve, and the hydraulic fluid will increase in temperature.

When a design requires that the piston can move freely, a 3-position valve of open-center type may be used (see Fig. 3.2). In such a valve the cylinder ports and the pump port are connected to the reservoir in the valve's mid-position.

In order to avoid system breakdown due to pressure overloads from excessive cylinder loads or a blocked positive displacement pump, convenient relief valving must be used.

Pilot-controlled check valves

Sometimes it is necessary to lock a cylinder from moving when the hydraulic supply pressure is dumped to zero. As shown in Fig. 3.3, this problem may be solved by using a pair of pilot-controlled check valves built into the cylinder connections between the directional control valve to the cylinder. The driving pressure is used as pilot control pressure for the check valve in the return connection to be opened.

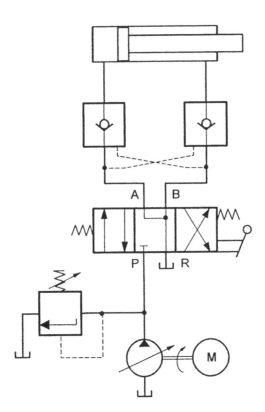

Fig. 3.3 Hydraulic circuit using pilot-controlled check valves to lock a piston

Motor speed control

A classical hydraulic control problem to be solved is the continuous control of piston or motor speed. Various solutions depending on application conditions and requirements are found. Here three frequently used systems will be briefly described (see Fig. 3.4). In all three cases the speed of a double-end piston in a double-acting cylinder is controlled in a continuous mode. The pressure medium supply, the power supply, is a pump with constant displacement protected against pressure overload by a pressure-relief valve. Note that in order to simplify the diagrams the directional control valve has been omitted.

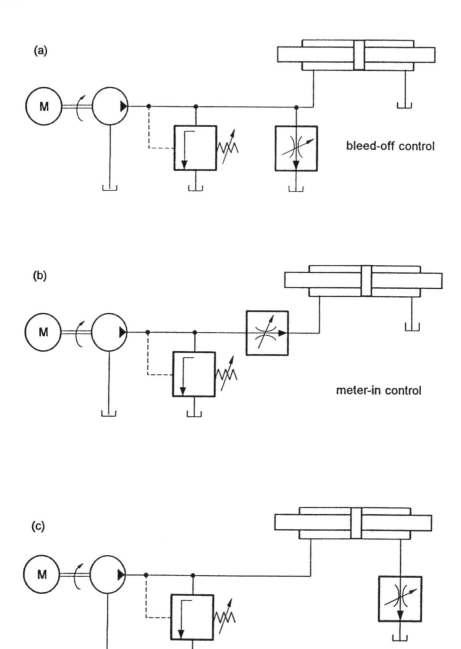

Fig. 3.4 Various ways of controlling piston speed

In Fig. 3.4a the pressure connection to the cylinder is equipped with a shunt connection (a bypass) to the reservoir. In this connection a flow control valve is built in. This arrangement is called **bleed-off**, because a part of the pump delivery bleeds off to the reservoir and is adjusted and controlled by the flow control valve. The rest of the pump delivery goes to the cylinder, and this flow rate determines the speed of the cylinder. The forward pressure on the piston is determined by the load. The power loss is determined by this pressure and the flow rate through the flow control valve. The bleed-off speed control is normally most convenient when the greater part of the pump delivery is used by the cylinder and the smaller part is bypassed.

Instead of a bypass in the pressure connection a flow control valve can be inserted in this connection, as seen in the system in Fig. 3.4b. The adjustment of the flow control valve directly controls the flow into the pressure side of the cylinder and thereby controls the piston speed. This arrangement is called a **meter-in** system. The portion of the pump delivery not used for the piston displacement returns to the reservoir as overflow through the pressure-relief valve. The pressure setting of this valve and the rate of overflow determine the power loss of the meter-in speed control system.

In Fig. 3.4c a system termed **meter-out** is illustrated. Now the pressure side of the cylinder is directly supplied from the pump. In the return connection from the cylinder to the reservoir a flow control valve is inserted. This valve controls the return flow rate from the cylinder and thereby the speed of the piston. The portion of the pump delivery not used for the piston displacement is overflow in the pressure-relief valve. The pressure on the pressure side of the cylinder is defined by the setting of the relief valve. The pressure on the return side is determined by the load on the piston. The main advantage of the meter-out speed control is its ability to control positive as well as negative loads, i.e. loads that either push or pull the motion of the piston. Like the meter-in system the meter-out system generates high power losses when only small fractions of the pump delivery are used for displacement of the cylinder. Compared with the bleed-off system, the meter-in and meter-out systems offer a high precision of speed control.

The systems in Fig. 3.4 are all shown with a linear actuator (cylinder), but rotary motors can just as well be used.

Back pressure valve

In some applications, e.g., in hydraulic presses and in feed speed control in drilling machines, where the load on a hydraulically driven piston suddenly disappears, a smooth operation can be achieved by using a so-called back pressure valve (see Fig. 3.5). In principle this valve is a pressure control valve inserted in the return connection of the cylinder and is controlled by using the return pressure, the back pressure, as pilot pressure. The back pressure valve, via the pilot's pressure-controlled throttling, will maintain a back pressure on the piston during load variations and thereby attempt to lock the piston between two columns of hydraulic fluid. This is sometimes termed a "locked circuit."

Sequence valve

The hydraulic circuit shown in Fig. 3.6 ensures a sequential motion of the pistons of cylinders *1* and *2*, first forward (to the right) and then backward (to the left). When flow via the directional control valve is pumped to the left chamber of cylinder *1*, the piston moves to the right, and when it meets the right bottom of the cylinder, the pressure will increase and the pilot-controlled pressure control valve *1* will open and allow flow to the left chamber of cylinder *2*. The piston of cylinder *2* then moves to the right. When the piston meets the right bottom of cylinder *2*, the motion stops.

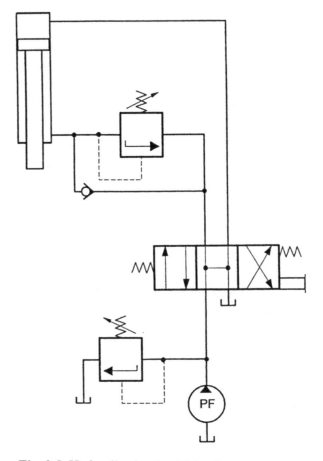

Fig. 3.5 Hydraulic circuit with back pressure valve

By shifting the directional control valve, the motions of the pistons will reverse, first the piston of cylinder *2*, then the piston of cylinder *1*.

The pressure control valves *1* and *2* in Fig. 3.6 are called sequence control valves.

Differential coupling

In some applications when using a double-acting cylinder with a single-end rod, the forward and reverse piston speeds are required to be equal. For such applications the differential coupling shown in Fig. 3.7 may be used. Here the full area, *A*, of the cylinder must be twice the ring area, *A/2*, at the piston rod side of the cylinder. The differential coupling may also be used for uneven forward and reverse piston speeds by an appropriate selection of the ratio of cross-sectional pressure areas of the two cylinder chambers.

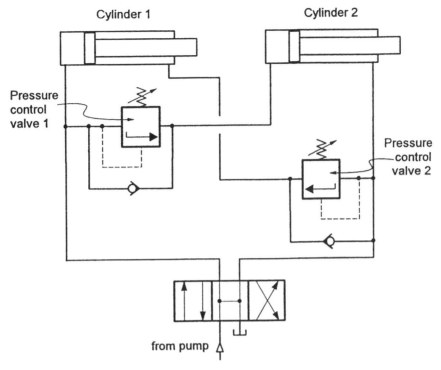

Fig. 3.6 Cylinder sequence circuit

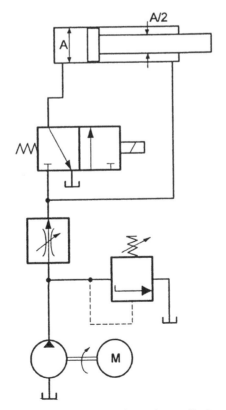

Fig. 3.7 Differential coupling of a double-acting cylinder with a single-end rod

Double-pump circuit

In many types of production machines two speeds are required: (1) a rapid traversing speed with a low supply of pressure and (2) a low speed with a high supply of pressure. In Fig. 3.8 a hydraulic system offering these features is shown. The system contains two pumps, a high-pressure (HP) pump with a low capacity and a low-pressure (LP) pump with a high capacity. The pumps are coupled in parallel to the system. When system pressure increases above the setting of the unloading valve, the LP pump flow is unloaded to the reservoir at zero pressure. The HP pump continues to supply the system with high pressure and low flow up to the pressure setting of the relief valve. This LP-HP pump coupling allows for a minimum rating of the power supply.

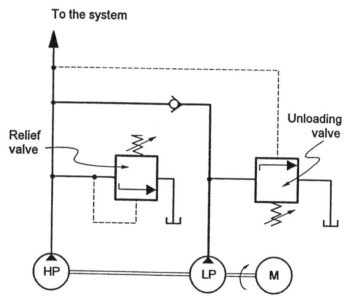

Fig. 3.8 Parallel coupling of a high-pressure (HP) and a low-pressure (LP) pump

Accumulator circuit

In Fig. 3.9 a hydraulic circuit using a **hydraulic accumulator** for energy storage is shown. A simple hydraulic accumulator (see Fig. 3.9, upper left) may consist of a spherical steel containment enclosing a flexible bladder filled with hydraulic pressure medium and connected with the hydraulic system. The space between the inside of the spherical containment and the outer side of the bladder is filled with pressurized gas (e.g., nitrogen).

The operation of the accumulator is as follows: when the pressure at the user site is below the setting of the unloading valve, the pump pumps pressure medium via the check valve to the system and the accumulator. When the pressure increases above the unloading valve setting, the valve opens and unloads the pump. Because of the check valve, the accumulator maintains the pressure at the user site above the setting of the unloading valve as long as the accumulator can deliver pressurized liquid. When this is not the case, the unloading valve closes and the pump starts up again. The circuit will function only when the closing pressure of the unloading valve is somewhat lower than its opening pressure.

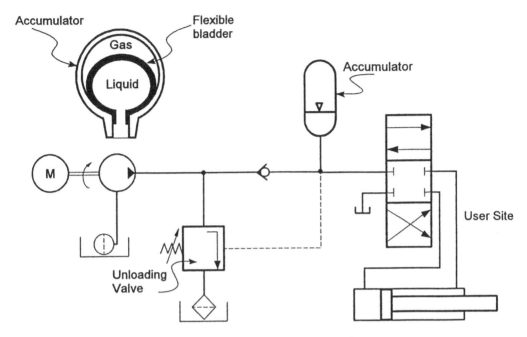

Fig. 3.9 Hydraulic circuit using an accumulator for energy storage

Hydrostatic transmission

A hydraulic system in which a pump drives a rotary motor is often termed as a **hydrostatic transmission**. Many different combinations of pumps and motors are used. Hydrostatic transmissions can be divided in two categories: open-circuit systems and closed circuit systems.

Open-circuit systems are characterized by having the return line for the pressure medium flow and the suction line of the pump connected to the reservoir.

Closed-circuit systems are characterized by having the pump ported directly to a rotary motor by directly connecting the pump pressure line to the motor pressure input port and the motor return line to the pump suction line.

In Fig. 3.10 the most often used type of hydrostatic transmission based on the closed-circuit principle is illustrated. A pump with variable capacity and capable of pumping in either direction is ported directly to a rotary, bidirectional motor with constant capacity.

The system in Fig. 3.10 is used for controlling the speed and direction of rotation of a hydraulic motor. In theory the system provides an infinitely variable transmission ratio, measured as the pump input speed over the motor output speed. The hydrostatic transmission is protected against pressure overload by the main pressure-relief valve and the combination of four check valves.

Via a combination of two check valves a booster pump is coupled to the low-pressure side of the closed circuit. The booster pressure (e.g., 8 bar) is limited by a pressure-relief valve.

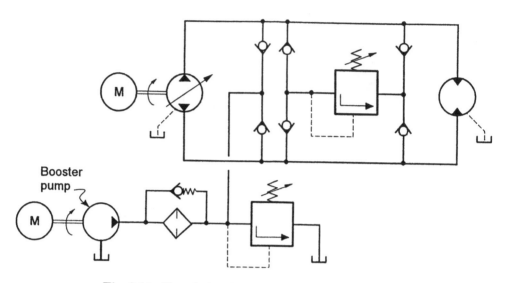

Fig. 3.10 Closed circuit for a hydrostatic transmission

The size of the booster pump is small compared with the main transmission pump, but the pump is still required in order to make up for internal leakage. Furthermore, the booster pump system prevents cavitation in the main pump, prevents overheating of the pressure medium due to hydraulic power losses and allows filtering of the fluid.

4. WATER AS THE HYDRAULIC MEDIUM

4.1 History of the fluid medium in hydraulics

The word **hydraulics** is defined as the science of the conveyance of water through pipes, etc.

Originally, water was the only available medium for hydraulics. In modern times, i.e., since the beginning of the 20th century, hydraulics as a fundamental discipline of mechanical engineering has come to include more than just water systems. It also deals with systems where not only water but oil as well as other liquids such as synthetic fluids are used as hydraulic media.

Furthermore, hydraulics is concerned not just with the conveyance of the liquid as such but also with the efficient transmission of pressure and power.

Hydraulic systems in mechanical engineering are most often **hydrostatic systems**, in which a liquid under an imposed pressure acts as a medium for pressure and power transmission.

The use of water as the first pressure medium in hydraulic systems had some immediate advantages. A primary one is that large quantities of water can flow in short time intervals, i.e., at high flow rates, through pipes with relatively small cross-sections without large pressure losses. This is due to the **low viscosity of water**.

A further advantage of using water is its **availability**, its **low cost** and the fact that water in the event of **leakage eventually evaporates without leaving greasy** or **dirty residues**. Finally, water does **not present a fire hazard** and it is **non-polluting**.

The primary disadvantages of water are its tendency to promote **corrosion** of normal steel and iron, its **lack of sufficient lubrication** between moving parts and a **somewhat limited range of working temperatures**.

To avoid some of the disadvantages, oil-in-water emulsions with 1-2% mineral oil are used in hydraulic systems, for example, for heavy forging presses. The emulsion creates an oil film protection reducing corrosion and offering some lubrication for the moving parts in valves, pumps and cylinders. Before emulsification the water has to be hardened to a pH value of approximately 8.

The availability of mineral oil and lubricating oils at the beginning of the 20th century, when the automobile industry was developing, made them a convenient pressure medium for hydraulic systems. Oil systems permit the use of high-speed piston pumps and sliding spool control valves. This was especially advantageous for smaller hydraulic plants. For very large, existing systems such as hydraulic presses using high-pressure (>300 bar) and seat valves, the only practical fluid is still water or oil-water emulsions. Most hydraulic systems today are mostly mineral oil-based. The practical need for systems to operate under restricted conditions, such as under extreme temperatures or pressures, or without risk of fire or environmental pollution, has led to the development of synthetic and/or water-based liquids as pressure media.

Over the last few decades several such new types of hydraulics fluids have been marketed. A liquid that is satisfactory as a hydraulic fluid in one application or system, however, may be completely unsuitable in another application or system. The requirements of a hydraulic fluid for a specific application will depend upon the specific system design and the operational conditions. Since there is no fluid possessing all properties to be ideal in all of the applications, the selection of a specific fluid for a given application will require compromises among properties to obtain the most suitable hydraulic fluid.

There are no analytical methods for the selection of a hydraulic fluid for a specific application. The selection must be made based upon a careful definition of the application requirements, previous experience and professional consultation with recognized hydraulic fluid suppliers.

The international and national standardization bodies (e.g., ANSI. ASTM, BS, DIN, ISO) have classified the basic types of hydraulic fluids in accordance with their composition and characteristics. These fluids have been coded with a string of capital letters.

In Figs. 4.1 and 4.2 (ref. 15) an overview of the basic types of hydraulic fluids is given, including a summary of their composition and main characteristics. In Fig. 4.1 the codes and descriptions for hydraulic mineral oils range from the simplest hydraulic oil (HH) to more sophisticated types with properties based upon various additives (HL, HLP, etc.).

ISO Code	Description
HH	Standard mineral oil without additives.
HL	Mineral oil with improved corrosion, anti-oxidation and temperature/viscosity properties. The selected viscosity classes range from 10 cSt to 100 cSt at 40°C.
HLP, HM	Mineral oil with improved corrosion, anti-oxidation, tempera-ture/viscosity and anti-wear properties.
HV	Mineral oil with low temperature effect on viscosity, usable for hydraulic systems in Arctic climates (high-viscosity index, VI)
HLPD	Mineral oil with additives that decrease stick-slip effect and emulge entrained water

Fig. 4.1 Mineral oils (petroleum-based fluids)

In Fig. 4.2 the codes and descriptions for hydraulic fluids classified as being fire-resistant are listed. There are two groups: one group is water-free and based upon synthetic fluids (HFD); the other group is based upon some content of water in emulsions or in solutions (HFA, HFB and HFC). Again, the codes classify the fluids according to various properties.

In recent years there has been an increasing interest in using hydraulic fluid that does not pose any risk as far as polluting the environment or the milieu around the hydraulic system by leakage of oil or waste. In Fig. 4.3 three special types of hydraulic fluids are classified according to their environmental friendliness.

In the above description nothing has been said about the cost of the different types of hydraulic fluids nor their availability. Recent experience shows that costs may vary with a factor of approximately 30 from the cheapest fluid (a mineral oil) to the most expensive fluid (an ester). Note that the price of mineral oil exceeds the price of pure tap water by a factor of 200! (Fig. 4.16). The possible unavailability at the user site of some hydraulic fluids—especially the more expensive ones—is another factor that must be considered.

ISO Code	Description
HFA	Oil-in-water emulsions (>80% water).
HFAE	Oil-in-water emulsions with anti-wear additives. Can be discharged to the drain (sewer).
HFAS	Aqueous solutions. Less aggressive to waste water than HFAE. Applied, for example, in coal mines and steel plants.
HFB	Water-in-oil emulsions. Flammable ingredients max. 60%. In some applications considered to be fireproof.
HFC	Aqueous polymer solutions. Polyglycol-water solutions in 6 classes of viscosity. Used, for example, in machinery for coal mines and diecasting machines.
HFD	These fluids are all water free and are synthesized on the basis of phosphate esters (HFDR), chlorinated and fluorinated carbons (HFDT) and other organic components (HFDU).

Fig. 4.2 Fire-resistant hydraulic fluids

ISO Code	Description
HPG	Polyglycol. Good lubrication and corrosive protection. Viscosity classification comparable to mineral oils.
HTG	Vegetable oil, such as rapeseed oil. Only few classes of viscosity.
HE	Synthetic ester. Viscosity classification comparable to mineral oils.

Fig. 4.3 Environmentally friendly hydraulic fluids

4.2 Water hydraulics

The future of hydraulic fluids will be greatly influenced by future developments of hydraulic components and systems as well as requirements from new applications. Considerations will focus upon the environmental conditions under which the hydraulic system must operate and upon those fields where the requirements are increased because of demands for more efficient and automatic power transmission and precise control of forces, pressures and velocities (flow rates).

With this perspective in mind the Danish company Danfoss, after more than 5 years of research and development efforts, has recently (1994) initiated the marketing of a complete new product range in water hydraulics based upon the use of **pure tap water** as the pressure medium. This range of **Nessie®** products includes hydraulic motors, pumps, valves, power packs and accessories and opens up a whole new series of applications.

The use of pure tap water introduces some important advantages over the use of other types of fluids, e.g., mineral oils and synthetic fluids:

- Pure tap water is **inexpensive, leakage** and **waste pose no pollution problems** and there are no **disposal difficulties**.
- Pure tap water does **not present fire risks** and is **safe** to use in hydraulic control for almost any type of process and machine.
- Pure tap water is **widely available** and does not consume a **significant amount of energy in its production**.

Water hydraulics is an old idea or concept, and the use of water as a pressure medium offering the above-mentioned advantages such as safety against pollution and fire risks and low initial and operating costs has long been recognized. But until recently, industry has widely believed that high-pressure water hydraulics for power control would not be possible.

Since the early days of hydraulics, the use of water as a hydraulic fluid has lead to insurmountable barriers such as poor lubrication between moving parts and destructive corrosion of vital parts in the hydraulic components. Also, the low viscosity of water has meant that clearances and tolerances between moving and stationary parts in the water hydraulic components have had to be kept very small in order to prevent excessive leakage and to maintain high system pressure. These experiences were, of course, based on the use of traditional materials, design techniques, technology.

The advent of the Nessie® concept has revolutionized the long-standing problem of how to use pure tap water as a hydraulic medium, given the inherent properties of water. What design principles and what materials may then be used in order to design and develop water hydraulic components without lubrication and corrosion problems and without excessive leakage so that the function becomes failsafe?

The answer to these questions has been the development of Nessie® water hydraulic components.

4.3 Physical properties of hydraulic fluids

This section defines and describes certain physical properties of fundamental importance to the performance of hydraulic systems. First, basic definitions of properties of hydraulic fluids are given. Then some functional relationships are illustrated graphically, and numerically typical values of the properties for the fluids pure water and mineral oil are given for illustration and comparison.

Density

The mass **density** ρ is defined as mass m per unit volume V:

$$\rho = \frac{m}{V}$$

The **weight density** γ is defined as the weight F per unit of volume V:

$$\gamma = \frac{F}{V} = \rho \cdot g$$

where g is the acceleration of gravity, $g = 9.81$ m/sec².

The **specific gravity** S_G is defined as the ratio of the mass (or weight) density of a material (a fluid or a solid) at a specified temperature t over the mass (or weight) density of water at the same temperature.

$$S_G = \frac{\rho}{\rho_w} = \frac{\gamma}{\gamma_w}$$

where ρ_w and γ_w are the mass density and weight density, respectively, of water at temperature t.

Mass is independent of pressure and temperature. Mass density, however, depends on pressure as well as temperature. In Figs. 4.4 and Fig. 4.5 the mass density ρ as a function of temperature and pressure is shown for mineral oil and water, respectively.

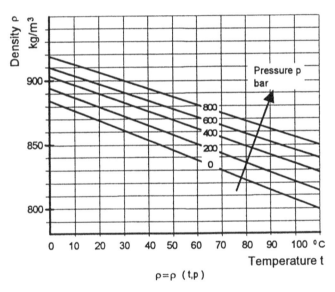

Fig. 4.4 Mass density ρ of a mineral oil as function of temperature t and pressure p

The mass density of hydraulic fluids is often referred to at the temperature 15°C ($\approx$60°F) and the atmospheric pressure 1 bar (absolute pressure).

The density $\rho(t)$ as a function of temperature t can be approximated by

$$\rho(t) \approx \rho(@15^0 C)\left[1 - \alpha(t - 15^0 C)\right]$$

where the cubic expansion coefficient α for a typical **mineral oil** is 0.00067 °C⁻¹ and for **water** is 0.00018 °C⁻¹.

Knowing the weight F of a charge of the fluid, the change ΔV of the volume by cooling or heating can be derived by

$$DV = \left(\frac{1}{g(t)} - \frac{1}{g(@15^0 C)}\right) \cdot F$$

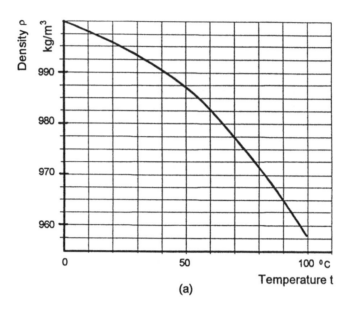

(a)

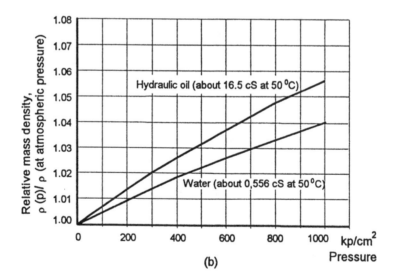

(b)

Fig. 4.5 (a) Mass density ρ of water as function of temperature t at atmospheric pressure

(b) Relative mass density $\rho(p)/\rho$ (at atmospheric pressure) as function of pressure p

The density $\rho(p)$ as a function of pressure p, assuming isothermal compression, can be approximated by

$$\rho(p) \approx \rho(@15°C \text{ and } 1 \text{ bar, abs.}) \left[1 + \frac{1}{\beta_T} p \right]$$

where the coefficient β_T is called the **isothermal bulk modulus (compression modulus)**. Sometimes the reciprocal of β_T is considered; it termed the **compressibility**:

$$c_T = \frac{1}{\beta_T}$$

It can be shown that the quantity β_T is the change in pressure Δp divided by the corresponding fractional change in volume at constant temperature

$$\beta_T \approx \frac{\Delta p}{\dfrac{\Delta V}{V}}\bigg]_{t \approx \text{constant}}$$

The bulk modulus β_T (and the compressibility c_T) varies with temperature and pressure. In practical computations, however, average values are normally used.

Practical values of the bulk modulus β_T for mineral oil lie in the interval $(1.33...1.54)\cdot 10^4$ bar.

A practical value of the bulk modulus β_T for water is $2.1\cdot 10^4$ bar.

The dependence of bulk modulus β_T on pressure and temperature for a mineral oil is shown in Fig. 4.6.

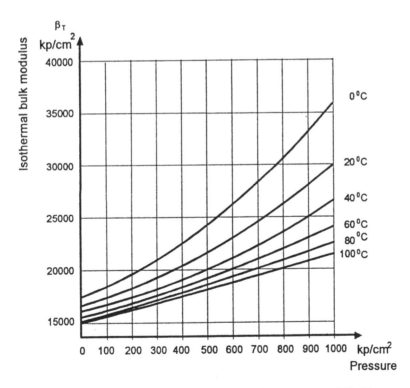

Fig. 4.6 Isothermal bulk modulus for mineral oil (from Dow and Fink's equation)

When the pressure/compression process in the hydraulic fluid is assumed to be isentropic/adiabatic, a little different bulk modulus, the **isentropic bulk modulus**

(compression modulus) β_S, is used. The isothermal and the isentropic bulk moduli are related by

$$\frac{\beta_S}{\beta_T} = k$$

where k is the ratio of the specific heats at constant pressure and constant volume. Typical values for β_S are $(1.0–1.6)\cdot10^4$ bar for mineral oil and $2.4\cdot10^4$ bar for water with no suspended air or vapor bubbles present.

Note that for engineering purposes the isentropic bulk modulus β_S is normally used.

Effective bulk modulus

The elasticity of the fluid in a hydraulic system and the elasticity of the walls of components (e.g., cylinders, pipes, valves and, in particular, flexible hoses) adds up to a total compressibility of the whole hydraulic system. The relation between this compressibility and the acceleration of load masses frequently causes resonances due to the inherently poor damping in most hydraulic systems. This source of possible hydraulic vibrations usually limits the performance of the hydraulic system when requirements of good dynamic behavior are specified.

It should be noted that the presence of suspended air bubbles (or bubbles from another gas) in the hydraulic fluid may drastically increase the compressibility (reduce the bulk modulus) of the fluid, especially in the lower pressure range (<50 bar). This effect is illustrated in Fig. 4.7, where the ratio of the theoretical bulk modulus β_v for a pure liquid and the effective bulk modulus β_e for a liquid containing a certain volumetric percentage of free air bubbles is shown as a function of pressure. As an example, oil <1% (volume) of free air in mineral oil may reduce the bulk modulus β_S 50%!

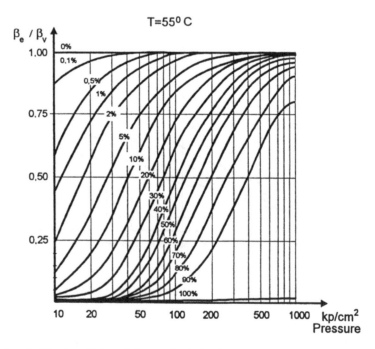

Fig 4.7 Ratio of effective (β_e) and theoretical (β_v) bulk modulus for a liquid as a function of pressure p and vol. % of free air at atmospheric pressure

Although the hydraulic fluid contributes most to the amount of the total compressibility of a hydraulic system, the contributions from other system elements may be of significance as well in specific applications. Therefore, for engineering computational purposes, the concept of **effective bulk modulus** β_e has to be introduced. In this concept the compressibilities of the various elements in a system are all lumped together in a single quantity, the effective bulk modulus β_e, primarily isentropic in nature.

An analytic computation of the effective bulk modulus is possible, in theory, but rather complex to carry out. In most cases a rule of thumb based on previous experiences is used to estimate a practical value for β_e. For a hydraulic system such a rule suggests a value for β_e equal to the half of the theoretical value stated above.

Solution of air in the hydraulic fluid

The dissolution of a gas in a fluid is usually expressed by using Bunsen's absorption coefficient b. This coefficient is defined as the volume of gas, measured at 0°C and at atmospheric pressure, that will dissolve in a unit volume of the hydraulic fluid at a partial pressure equal to the atmospheric pressure. Bunsen's absorption coefficient b is approx. 0.09 for mineral oil and 0.02 for water (see Fig. 4.8).

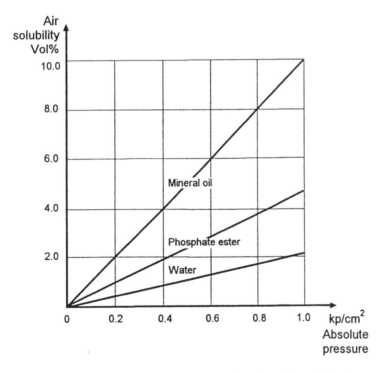

Fig. 4.8 Air solubility for some pressure media as function of absolute pressure.

The amount of air that will be dissolved in the hydraulic fluid at high pressures taken in a unit volume of the fluid and measured at the prevailing pressure will be approximately constant at constant temperature. As indicated by the Bunsen coefficient the content of air will then be approximately vol. 9% for mineral oil and vol. 2% for water.

The air dissolved in the hydraulic fluid will change the mechanical properties to only a negligible degree. The presence of air in the fluid as free air bubbles or in dissolved state depends largely upon the pressure conditions. A saturated fluid-air solution will liberate free air bubbles when the pressure decreases, e.g., in the suction line of the pump or in a throttling process in a valve. In fact, the liberation of free air is considered as a kind of cavitation, and it may disturb the correct function of the hydraulic system.

Cavitation

In hydraulic systems the term **cavitation** can be defined as the formation and collapse of cavities within the fluid. The cavities are air or fluid vapor bubbles. The collapse of these bubbles (implosion) takes place at regions where the pressure suddenly rises, and it is within such zones that cavitation may cause damage to the surrounding component walls or the moving members of valves, pumps, etc.

Suspended air bubbles may be present in the fluid from the fluid entrance through the pump suction line or may be liberated from the fluid when a depressurization occurs (see above). The hydraulic system design should minimize the occurrence of free air bubbles.

Fluid vapor bubbles will arise when the pressure is below the vapor pressure, i.e., the pressure at which vaporization initiates. The vapor pressure is dependent on the temperature of the fluid. The lower the temperature, the lower the vapor pressure will be, which means that pressure has to be reduced considerably before the fluid will boil.

The vapor pressure as a function of temperature for some hydraulic fluids is shown in Fig. 4.9.

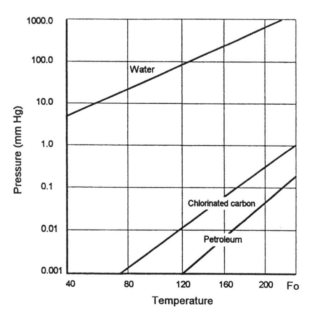

Fig. 4.9 Vapor pressure of some hydraulic fluids (ref. 3)

Cavitation may ruin the function of the hydraulic system and may destroy or considerably shorten the life of the components. As can be seen from Fig. 4.9, water may be used up to a temperature of only 122°F (= 50°C) in order to avoid the risk of cavitation due to water vapor.

Viscosity

Viscosity is a measure of the internal friction in a fluid when a layer of the fluid is moved in relation to another layer and thereby causes friction between the molecules of the fluid when they are moving with uneven velocities. Viscosity is one of the most important properties of a hydraulic fluid and has a great influence on the performance of the hydraulic system.

If the viscosity is too high, component efficiencies decrease because of the power loss (pressure drop) to overcome fluid friction during fluid flow. On the other hand, if the viscosity is too low, external as well as internal leakage will increase and thereby cause power loss. Furthermore, a certain optimal value of viscosity is vital for hydrodynamic lubrication.

The selection of hydraulic components for their functional design principle, dimensions, tolerances and clearances must therefore match the viscosity of the hydraulic fluid in order for the system to exhibit an optimal system performance.

When the temperature increases or the pressure decreases, the intermolecular forces decrease and thereby the viscosity of the fluid will normally decrease. A precise statement of the viscosity therefore requires a reference to temperature and to a reference pressure.

The following definition for viscosity is attributable to Isaac Newton and is based upon the assumption that the velocity distribution between two large parallel plates at a small distance B apart is linear. The space between the plates is filled with a fluid. A tangential force F must be acting upon the upper plate when it is moving with a constant velocity V (see Fig. 4.10).

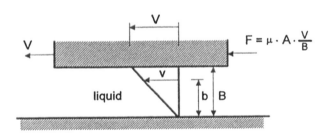

Fig. 4.10 Definition of dynamic viscosity μ

The fluid in contact with the upper plate adheres to the plate and moves with the velocity V. Likewise, the fluid adheres to the fixed plate and has zero velocity. Experiences has shown that force F varies proportionally with the area A of the plates and the shear stress τ in the fluid:

$$F = A \cdot \tau$$

The rate of shear, the velocity gradient, G is expressed by

$$G = \frac{v}{b} = \frac{V}{B}$$

By introducing a coefficient μ, **the absolute or dynamic viscosity**, Newton derived the viscosity law by

$$F = \mu \cdot A \cdot G = \mu \cdot A \cdot \frac{V}{B}$$

where

$$\mu = \frac{\dfrac{F}{A}}{\dfrac{V}{B}} = \frac{\text{shear stress}}{\text{velocity gradient}}$$

The concept of absolute or dynamic viscosity is used mostly by physicists. Normally in engineering practice the quantity ν, **kinematic viscosity,** is used.

The kinematic viscosity is defined by

$$\nu = \frac{\mu}{\rho}$$

where ρ is the mass density.

For dimensions of viscosity and conversion between physical units, refer to Appendix A.

Fluids behaving in accordance with the above law are called Newtonian fluids.

Because in practice it has not been possible through a direct method to measure the dynamic or kinematic viscosity, a whole series of indirect measurement principles have been developed, and especially used in the oil industries. In Appendix B a conversion table for the most-used empirical units is presented.

Viscosity-temperature relationship

Some liquids such as water exhibit little change in viscosity due to temperature changes.

The temperature-induced change in viscosity of mineral oils is normally so significant that the choice of oil type for a given hydraulic system requires a serious evaluation of whether the viscosity changes that the selected oil will exhibit are acceptable or not.

In Fig. 4.11 the viscosity-temperature relation for **water** is shown. As can be seen from Fig. 4.11, the viscosity (in centipoise) of water will decrease with a factor of approximately 3 when the water temperature varies from 3°C to 50°C.

In Fig. 4.12 the viscosity-temperature relationship for a typical **hydraulic mineral oil** is shown. The viscosity (in centistoke) of the oil will decrease with a factor of approximately 6.0 when the oil temperature varies from 30°C to 70°C.

Even though the relationships in Fig. 4.11 and Fig. 4.12 are relatively easy to determine by experiments, there seems to be no fundamental mathematical relation for an exhaustive description of the influence of temperature variation on corresponding viscosity for liquids.

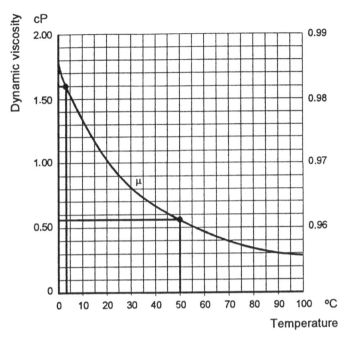

Fig. 4.11 Dynamic viscosity of water as function of temperature at atmospheric pressure

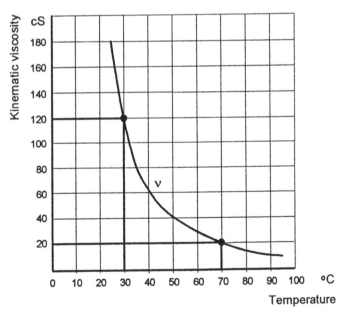

Fig. 4.12 Kinematic viscosity of a mineral oil as function of temperature at atmospheric pressure (ref. 19)

However, for most types of mineral oils the following formula (Walther's formula) has led to a graphic linear relationship for viscosity as function of temperature:

$$\log_{10}\log_{10}(v + k) = A + B \cdot \log_{10} T$$

where

$$\nu = \text{kinematic viscosity in cS}$$
$$T = \text{absolute temperature in °K}$$
$$A \text{ and } B = \text{constants specific for the given oil}$$
$$k = \text{a universal constant (except for low values of } \nu)$$

The above equation is the basis for the ASTM's (American Society for Testing and Materials') double-logarithmic viscosity diagrams. In these diagrams a straight line is drawn between two measured data points. A typical example of such a diagram is shown in Fig. 4.13.

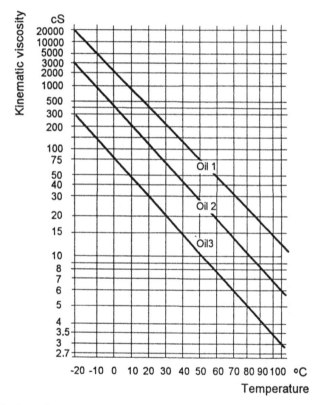

Fig. 4.13 Double-logarithmic diagram of viscosity for some mineral oils at atmospheric pressure.

Viscosity-pressure relation

The viscosity of all liquids will normally increase with the pressure. For hydraulic low-pressure systems the effect is negligible. For pressures above 200 bar the change in viscosity becomes significant. The rate of change increases with decreasing temperature. For a pressure of about 350 bar the viscosity of a typical mineral oil will be double the value of the viscosity as a function of the pressure. In the pressure range 0-1000 bar the following approximation can be used:

$$\mu = \mu_o \cdot e^{\kappa \cdot p}$$

where

μ = dynamic viscosity at pressure p

μ_0 = dynamic viscosity at the atmospheric pressure

κ = pressure coefficient of viscosity depending of the fluid and range of pressure temperature (a typical value for mineral oil is $\kappa \approx 10^{-3}$ bar^{-1})

As the viscosity of a fluid depends both on the pressure and the temperature, it would be desirable to have just one integrated analytic relationship expressing this dependence. So far, however, this has not yet been possible and its determination is still be an object of research.

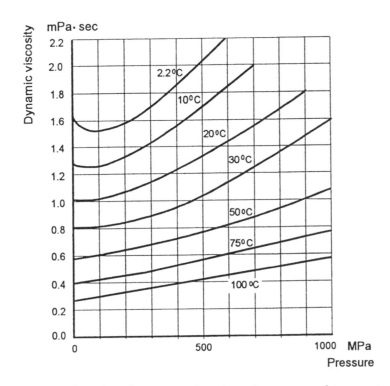

Fig. 4.14 Absolute viscosity of water as a function of pressure and temperature (ref. 8)

In Fig. 4.14 the dependence of water viscosity on pressure is illustrated. The diagram also shows the influence of temperature.

Likewise, in Fig 4.15 the dependence of mineral oil viscosity on temperature is illustrated and the influence of the pressure is shown.

4.4 Tap water as the hydraulic pressure medium

Applications using clean water hydraulic systems such as the Nessie® technology are still in their infancy. Therefore, in the following sections we will focus primarily on the experiences gained so far from numerous laboratory and practical field tests and from those using the emerging applications of Nessie® technology.

The use of **tap water** as a hydraulic pressure medium naturally raises many technical questions to be answered.

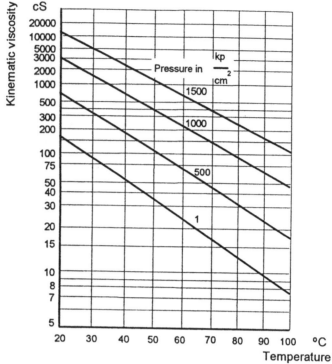

Fig. 4.15 The kinematic viscosity of a mineral oil as a function of temperature and pressure
(ref. 19)

Basic properties of tap water

Tap water, although it is nearly universally available, is by no means of a uniform or standard quality. The homogeneity and composition of water are very dependent on geographical locations. Not only from country to country but also from region to region there is a considerable variance in quality of water. In some cases tap water may be pumped up from a boring or well; in other cases, it is extracted from lakes and rivers. In the latter case, the water has to go through a complex handling process. Quite often chlorine is used in order to kill bacteria and other microorganisms in order to make tap water potable (i.e., drinking water).

For the time being, the most comprehensive effort to define the quality requirements for drinking water in Europe has been set forth in the European Union Directive 80/778/EEC (1980).

In this directive many parameters and their limiting values as to which substances the drinking water may contain have been specified. Some of the parameters specify the microbiological requirements for drinking water. Fortunately, only a few of the above parameters play a significant role for the function, lifetime and safety of hydraulic systems using tap water as a pressure medium.

Those parameters from the directive that are of importance to Nessie® water hydraulic technology are briefly commented on in the following paragraphs:

A. **Hydrogen ion** concentration expressed as the pH value of water. The EU directive specifies pH values between 6.5 and 8.5, which too is acceptable for Nessie components. Lower pH values can cause **corrosion** problems in a tap water hydraulic system.

B. Contents of **chloride ions** (Cl^{-1}). The directive specifies a maximum value of 25 mg/l. In practice much higher values may be found. When a chloride content exceeds 200 mg/l it may cause crevice **corrosion** problems also on stainless steel.

Free chlorine (Cl_2) must not be present in drinking water. It may also increase the risk of corrosion.

C. **Hardness** of water. The hardness expresses the amounts of magnesium (Mg) and calcium (Ca) in water. The hardness of water is calculated from the following formula:

$$\frac{Ca\,(mg/l)}{7.13} + \frac{Mg\,(mg/l)}{4.35} = °d \qquad (\text{"German" degree of hardness})$$

A conversion table of German (and Danish), English, French, and U.S. hardness measurement units is found in Appendix C.

In the EU directive no maximum value is given. Experiences from use of Nessie® systems show that no problems will occur when the hardness lies in the interval 5-10 °d. High hardness causes increased risk of formation of lime deposits and eventual clogging of narrow passages or crevices in the systems.

Lime is deposited electrolytically at cathode areas and on surfaces with a temperature of more than 50-60 °C. When the flow through components, pipes, etc., is high enough, the lime will detach and be transported by the water to the filter. If the water has a high content of lime it can be necessary to remove it with carbon oxide.

Corrosion in tap water hydraulic systems can be controlled by proper selection of the applied materials in the hydraulic components and by maintaining the quality of the tap water (see points A and B above). The preferred material is stainless steel. Other materials—depending on the specific application—such as bronze, brass, polymers, anodized alumina, and ceramics may also be used. In some cases normal steel with various coatings or chromium-plated steel may be used.

It is to be noted that water is electrically conductive and may act as an electrolyte when impurities (such as detergents) or certain additives are present. In such cases electrolytic corrosion may arise. Therefore metallic materials to be used in connection with the water pressure medium should comply with the electrochemical series. For this reason untreated aluminum and zinc should not be used, whereas copper alloys and nickel chrome steels are preferable.

Bacteria in tap water hydraulic systems

In connection with water hydraulic liquids of type HFA and HFB, bacterial growth has often been a problem. Typically the growth of bacteria, fungi, molds, and other microorganisms results in clogged filters and bad smells. For HFA fluids this problem has traditionally been dealt with through the addition of additives, which can create new problems.

In water hydraulic systems bacteria and fungi may grow and be damaging in two other ways:

(1) The bacteria and fungi can pollute the environment via leakage
(2) A biocoating on the inner side of the hydraulic components may arise and may thereby introduce corrosion.

Microorganisms inhabit a wide variety of aqueous environments. In order for microorganisms to survive and reproduce in water, certain conditions must be present. One condition is the presence of nutrients (salts, proteins, etc.) in the water. Another condition is a suitable environmental temperature, typically in the range of 5-60 °C.

The Nessie® technology uses ordinary tap water as a pressure medium, without the addition of any antibacterial or other additives. Therefore, nutrients for bacteria and fungi must be prevented from entering the tap water that is to be used. To ensure that a hygienic level is maintained, the following set of precautions has to be taken:

- The air breather (air-filter) of the reservoir/tank should have a filter fineness of better than 3 μm (absolute).
- The tank lid must fit absolutely tight.
- Refilling of water must be done via a filter (return filter).
- Quick coupling connections must be kept absolutely clean.
- Ingression of "foreign material," for example, via packings at piston rods and motor shafts, must be avoided.

The content of bacteria and other microorganisms in the water is most often measured by expressing the total count of bacteria per ml of the water. The EU directive mentioned on p. 52 specifies only recommended values for clean drinking water such as 10 bacteria/ml @ 37°C and 100 bacteria/ml @ 22°C. There are, for the time being, no EU directives that specify maximum limits.

In any closed or open water system, depending on temperature conditions, there will be inputs from outside of microorganisms and nutrients, etc., and some growth in the amount of such organisms will happen. However, there are so far no rules or specifications on critical values not to be exceeded.

Early experiences with Nessie® water hydraulic systems have shown that the growth described stabilizes rather soon at a level that does not impair the function of the Nessie® system, and no traces of harmful bacteria have been found as long as a stable and acceptable hygienic level was maintained. Depending on hygienic requirements, it can become necessary to flush the water hydraulic system with a cleaning agent and then refill it with tap water.

In Fig. 4.16 (p. 57) a comparison of the various characteristic properties of water and other frequently used hydraulic fluids is presented.

In short the advantages of using pure tap water over other liquids as a pressure medium for hydraulic system are illustrated in the matrix in Fig. 4.17. The properties of environmental impact and fire hazard are compared for various pressure media.

Example 4.1

A typical hydraulic mineral oil has a cubic expansion coefficient $\alpha = 0.00065$ °C^{-1}. When temperature increases from 15°C to 50°C, the relative change of mass density becomes

$$\left| \frac{\rho(t) - \rho(@15°C)}{\rho(@15°C)} \right| = 0.00065 \cdot 35 \cdot 100\% = 2.28\%$$

Example 4.2

In a hydraulic cylinder made of steel and having the dimensions indicated in Fig. 4.18, the hydraulic pressure changes from p_1 to p_2.

Assuming that the cylinder is infinitely stiff, the piston will change its position, Δh_1, due to the compressibility of the hydraulic medium when the pressure changes.

The effective bulk modulus, β_e, of the pressure medium varies with pressure, as shown in Fig. 4.18. The position change Δh_1 can be derived as follows:

By definition

$$\beta_e = \frac{\Delta p}{\dfrac{\Delta V}{V}} \qquad \text{(see section 4.3)}$$

In the pressure range considered here, β_e varies approximately linearly with the variation in pressure. The variation in volume, ΔV, can be expressed by

$$\Delta V = V \cdot \frac{1}{2} \cdot \left(\frac{1}{\beta_{e1}} + \frac{1}{\beta_{e2}} \right) (p_2 - p_1)$$

where

$$\Delta h_1 = \frac{\Delta V}{L} = \frac{L}{2} \cdot \left(\frac{1}{\beta_{e1}} + \frac{1}{\beta_{e2}} \right) (p_2 - p_1)$$

With the values $p_1 = 150$ kp/cm², $p_2 = 300$ kp/cm², L = 40 cm, $\beta_{e1} = 1/65 \cdot 10^6$ kp/cm² and $\beta_{e2} = 1/57 \cdot 10^6$ kp/cm² (the latter two values are determined by using Fig. 4.18), Δh_1 is derived by

$$\Delta h_1 = \frac{40}{2} \cdot (65 \cdot 10^{-6} + 57 \cdot 10^{-6}) \cdot (300 - 150) = 0.366 \text{ cm} = 3.66 \text{ mm.}$$

Note that Δh_1 is independent of the cylinder diameter D.

Example 4.3

Through a 10-m-long pipe a power of 10 kW with a degree of efficiency $\eta = 0.97$ has to be transmitted to a user. At the user site a pressure $p_2 = 210$ kp/cm² is required. The mass density ρ of the pressure medium is 0.9 kg/dm³ and the kinematic viscosity $v = 30$ cS. The pump supply pressure p_1 and the pipe diameter have to be determined.

The power at user site is expressed by

$$P_2 = p_2 \cdot Q_2$$

where p_2 is the pressure at the user site and Q_2 is the flow rate.

The power P_1 at the pump site is expressed by

$$P_1 = p_1 \cdot Q_1$$

where p_1 is the pressure at the pump site and Q_1 is the flow rate.

Using the degree of power efficiency the following relation holds:

$$P_2 = p_2 \cdot Q_2 = \eta \cdot p_1 \cdot Q_1$$

When $Q_1 = Q_2$ and $p_2 = 210$ kp/cm² then

$$p_1 = \frac{1}{\eta} \cdot p_2 = \frac{1}{0.97} \cdot 210 = 216.5 \text{ kp} / \text{cm}^2$$

$$Q_1 = \frac{P_1}{p_1} = \frac{10^4}{216.5 \cdot 9.81 \cdot 10^4} \text{ m}^3 / \text{sec} = 28.2 \text{ l} / \text{min}$$

Assuming laminar flow in the pipe, the Hagen-Poiseuille formula (section 1.3) gives:

$$D^4 = \frac{128 \cdot \mu \cdot L \cdot Q}{\pi (p_1 - p_2)} = \frac{128 \cdot 0.3 \cdot 10^{-4} \cdot 900 \cdot 10 \cdot 10 \cdot 10^3}{\pi \cdot 6.5 \cdot 9.81 \cdot 10^4 \cdot 216.5 \cdot 9.81 \cdot 10^4} = 0.82 \cdot 10^{-8} \text{ m}^4$$

$$D = 0.95 \cdot 10^{-2} \text{ m} = 9.5 \text{ mm}$$

Reynolds number (section 1.3):

$$R = \frac{v \cdot D}{v} = \frac{4 \cdot P_2}{\pi \cdot D \cdot p_2 \cdot v} = \frac{4 \cdot 10 \cdot 10^3 \cdot 0.97}{\pi \cdot 9.5 \cdot 10^{-3} \cdot 210 \cdot 9.81 \cdot 10^4 \cdot 0.3 \cdot 10^{-4}} = 2024$$

(considered to be laminar flow).

Example 4.4

The same pipe as in example 4.3 is now used for flow with a pressure medium (water) with a mass density $\rho = 1.0$ kg/dm³ and a kinematic viscosity $v = 1$ cS. The same pressure drop along the pipe is maintained. Assuming the flow is **turbulent,** the flow rate Q can be computed by

$$Q^{1.75} = \frac{D^{4,75}}{0.242 \cdot v^{0,25} \cdot \rho} \cdot \frac{p_1 - p_2}{L} \quad \text{(ref. section 1.3)}$$

$$= \frac{(0.95 \cdot 10^{-2})^{4.75}}{0.242 \cdot (10^{-6})^{0.25} \cdot 1000} \cdot \frac{6.5 \cdot 9.81 \cdot 10^4}{10}$$

$$= 5.641 \cdot 10^{-4} \text{ m}^3 / \text{sec} = 33 \text{ l} / \text{min}$$

The Reynolds number (section 1.3) becomes

$$R = \frac{v \cdot D}{v} = \frac{4 \cdot Q}{\pi \cdot v \cdot D} = \frac{4 \cdot 5.641 \cdot 10^{-4}}{\pi \cdot 10^{-6} \cdot 0.95 \cdot 10^{-2}} = 75600$$

Comparing the results with example 4.3, the flow rate has increased approximately 20%.

Liquid	Mineral oil HLP	HFA	HFC	HFD	Bio-oil (rapeseed) HTG	Water
Kinematic viscosity at 50°C [mm²/sec]	15 - 70	~ 1	20 - 70	15 - 70	32 - 46	0.55
Density at 15°C [g/cm³]	0.87 - 0,9	~ 1	~ 1.05	~ 1.05	0.93	1
Vapor pressure at 50°C [bar]	$1.0 \cdot 10^{-8}$	0.1	0.1 - 0.15	$<10^{-5}$	?	0.12
Compression modulus β_s [N/m²]	$1.0 - 1.6 \cdot 10^9$	$2.5 \cdot 10^9$	$3.5 \cdot 10^9$	$2.3 - 2.8 \cdot 10^9$	$1.85 \cdot 10^9$	$2.4 \cdot 10^9$
Speed of sound at 20°C [m/sec]	1300	?	?	?	?	1480
Thermal conductivity at 20°C [W/m·°C]	0.11 - 0,14	0.598	~ 0.3	~ 0.13	0.15 - 0.18	0.598
Specific heat at 20°C and constant pressure [kJ/kg·°C]	1.89	-	-	-	-	4.18
Max. working temperature range [°C]	-20 - 90	5 - 55	-30 - 65	0 - 150	-20 - 80	~ 3 - 50
Flash point [°C]	210	-	-	245	250 - 330	-
Ignition point [°C]	320 - 360	-	-	505	350 - 500	-
Corrosion protection	Good	Sufficient	Good	Good	Very Good	Poor
Environmental impact	High	High	High	High	Small	None
Relative costs for liquid [%]	100	10 -15	150 - 200	200 - 400	150 - 300	~ 0.02
Usage [%]	85	4	6	2	3	~ 0 (at present)

Fig. 4.16 Characteristics of water and other hydraulic fluids

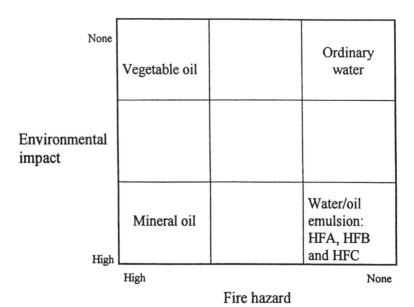

Fig. 4.17 The primary advantages of using ordinary water as a hydraulic pressure medium

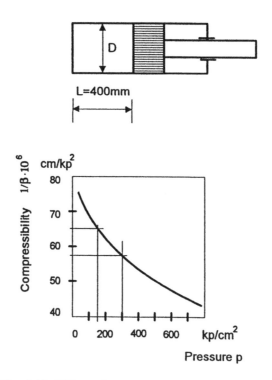

Fig. 4.18 Effect of pressure medium compressibility

5. HYDRAULIC PUMPS

5.1 Positive displacement principle

Hydraulic pumps can be divided into two classes: pumps that are suited for the transfer of fluids and pumps that are suited for the transfer of energy, i.e., fluid pressure applications.

The two classes of pumps are in many cases based upon the same mechanical principles but may differ in design details and dimensions. It is the latter class of pumps that is of interest in this context.

Another way of classifying hydraulic machines (i.e., pumps and motors) is to distinguish between "**positive**" and "**non-positive**" machines.

Non-positive pumps contain an impeller or a propeller that is used to generate a radical and/or an axial liquid flow. When an impeller is used, the mean flow in the pump has a radial velocity component. The pump is called a centrifugal pump. Analogously, when a propeller is used, the mean flow in the pump has an axial velocity component and the pump is called an axial pump. Non-positive pumps are rarely used in power hydraulics.

Positive pumps or more correctly, positive displacement pumps, are pumps in which fluid entering the pump chamber or chambers through the inlet is forced out again through the delivery outlet. Internal leakage is neglected. The fluid is positively displaced by the action of one or more moving members in the pump (see Fig. 5.1).

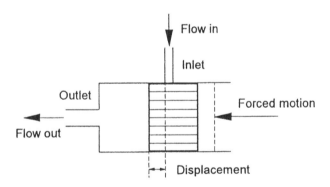

Fig. 5.1 Positive displacement principle

In order that the positive displacement principle can be applied to a continuous pumping of a fluid liquid or to the development of mechanical power by a motor, the idealized piston-cylinder unit discussed above (Fig. 5.1) must be slightly modified. Note that theoretically speaking a hydraulic motor is essentially a reversible pump.

In Fig. 5.2 a simple reciprocating single-piston pump is shown. When the piston is moved to the right, liquid is displaced and forced out through a check valve in the outlet. The check valve in the inlet pipe is closed due to a pressure rise in the cylinder chamber. When the piston is moved to the left, the pressure in the cylinder chamber is reduced, whereby the check valve in the outlet pipe closes and the check valve in the inlet pipe opens. The reduced pressure in the cylinder chamber during the stroke of the piston to the left allows the atmospheric pressure in the reservoir to force the liquid up into the cylinder chamber through the inlet pipe from the reservoir. The piston reciprocates by a crankshaft mechanism. Having a number of piston-cylinder units driven in parallel and sequentially timed helps to maintain a continuous pump flow rate.

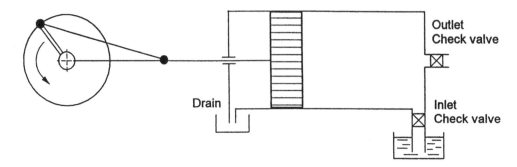

Fig. 5.2 A simple reciprocating pump

The piston-cylinder unit in Fig. 5.2 may operate as a motor if the two check valves are replaced by valves timed to operate in accordance with the motion of the piston. When the liquid under pressure enters the cylinder chamber through the open valve in the motor inlet (previously the pump outlet) and the valve in the motor outlet (previously the pump inlet) is closed, the piston is forced to move to the left. If the inertia in the crankshaft rotation is high enough, the piston returns after ending the stroke to the left if the valve in the motor inlet is closed and the valve in the motor outlet (to the reservoir) is opened, allowing the fluid to discharge to the reservoir. Having a number of piston-cylinder units working in parallel helps to maintain a continuous motor output rotation of and power transfer to the crankshaft.

Two important definitions can now be stated:

1. The displacement of a positive, hydraulic pump or motor equals the volume that is displaced per revolution of the driven shaft or driving shaft, respectively.

2. The delivery of a positive, hydraulic pump or motor is the rate at which liquid is discharged in units of volume per time.

Using the above definitions, the flow rate Q in volume units per time of a positive, hydraulic pump or motor is given by

$$Q = D \cdot n$$

where D is the displacement in volume units per revolution and the rotational speed n in revolutions per unit time.

In an ideal pump (or motor) the mechanical power input (power output) $T \cdot \omega$ equals the hydraulic power output (power input) $Q \cdot p$, where ω equals the angular velocity in radians per unit time, T equals the applied torque and p equals the hydraulic pressure. This is written as:

$$T \cdot \omega = Q \cdot p$$

In this context only positive displacement machines will be considered.

5.2 Stationary analysis of hydraulic machines

In stationary analyses and static dimensioning of hydraulic systems, dynamic relations, such as pump pulsations and kinematic variations in the displacement of pumps and motors, are neglected. The stationary behavior of hydraulic machines is characterized by the relationships among the four main variables: **torque, speed, flow rate** and **pressure**. The variables are evaluated by their averaged values. In the following considerations rotating machines are assumed. However, analogous expressions may easily be worked out for reciprocating machines.

Ideal stationary relationships

A hydraulic pump or motor can be considered as being loss free, i.e., it works with a 100% degree of efficiency (definition is given later). For an ideal pump (subscript P for pump) the **mechanical power input**, E_{inP}, is expressed by:

$$E_{inP} = T_P \cdot \omega_P$$

where T_P is the input torque on the pump shaft and ω_P is the angular velocity (in radians per time).

The **hydraulic power output**, E_{outP}, is expressed by

$$E_{outP} = p_P \cdot Q_P$$

where p_P is the output hydraulic pressure of the pump (the pump intake pressure taken as reference) and Q_P is the average output volume flow rate.

Analogous equations for an ideal motor (subscript M for motor) are as follows:

$$E_{inM} = p_M \cdot Q_M$$

$$E_{outM} = T_M \cdot \omega_M$$

Per definition

$$D = \frac{Q \cdot 2\pi}{\omega}$$

where D is the displacement in volume units of the pump or the motor per revolution and ω is the angular velocity of the motor or pump.

For ideal hydraulic machines the parameter D is sufficient for specifying the respective machine. But also for practical machines the parameter D is of utmost significance.

From the equations above the following relations can be derived:

$$T_P = \frac{Q_P}{\omega_P} \cdot p_P = \frac{D_P}{2\pi} \cdot p_P$$

$$T_M = \frac{Q_M}{\omega_M} \cdot p_M = \frac{D_M}{2\pi} \cdot p_M$$

In practice it is common to characterize a hydraulic motor by its displacement D_M. Hydraulic pumps are most often referred to by their delivery volume flow rate, Q_P, at a certain rpm, $n_P = \omega_P / 2\pi$. From the above D_P is determined by Q_P / n_P

Definition of efficiencies

In practice hydraulic pumps and motors can work properly only if they are made with some clearances or gaps between the moving parts. Furthermore inevitable tolerances of the parts in a motor or pump assembly will result from the manufacturing. Consequently the parts will deviate from the ideal specifications, whereby unwanted friction and gaps will occur. These relations will lead to power losses in pumps and motors and may cause non-ideal operational behavior.

The power losses are in practice identified in two ways:

1. During practical operational conditions the delivery rate of a pump decreases and the assumed flow rate of a motor increases when pressure increases, assuming the speed, ω_P or ω_M, is constant. This is caused by the increase of leakage through the clearances and gaps (discussed above) when the pressure is increasing.

   The power losses caused by the leakage are defined as the **volumetric losses**.

2. Further, during practical operational conditions the effective input torque on the input shaft of the pump must necessarily be larger than the theoretical torque T_P (see above), and the effective output torque on the output shaft of the motor must likewise necessarily be smaller than T_M (see above). This is due to internal friction and flow resistance in the hydraulic machines and is caused by mechanical as well as hydraulic relationships.

   The power losses caused by the friction and resistance are defined as the **mechanical losses**.

For a pump the volumetric losses can be expressed as follows:

$$Q_{eP} = Q_P - Q_{sP}$$

where

Q_{eP} = the effective, averaged pump delivery per unit time (volume flow rate).
Q_P = the theoretical, averaged pump delivery per unit time.
Q_{sP} = the averaged pump leakage flow rate and loss flow rate. The latter is caused by cavitation (free air and vapor).

Analogously, for a motor the volumetric losses are expressed by:

$$Q_{eM} = Q_M + Q_{sM}$$

Now a **volumetric efficiency**, η_v, may be defined in the following way:

For **a pump**: $\eta_{vP} = \dfrac{Q_P - Q_{sP}}{Q_P}$

For **a motor**: $\eta_{vM} = \dfrac{Q_M}{Q_M + Q_{sM}}$

The mechanical losses for a pump are expressed as follows:

$$T_{eP} = T_P + T_{sP}$$

where

T_{eP} = the effective, necessary torque on the pump shaft
T_P = the theoretical, necessary torque on the pump shaft
T_{sP} = the part of T_{eP} caused by the friction relations and flow resistances in the pump

Analogously, for a motor the mechanical losses are expressed by:

$$T_{eM} = T_M - T_s$$

Now a **mechanical efficiency**, η_m, may be defined in the following way:

For **a pump**: $\eta_{mP} = \dfrac{T_P}{T_P + T_{sP}}$

For **a motor**: $\eta_{mM} = \dfrac{T_M - T_{sM}}{T_M}$

A **total efficiency**, η_t, for pumps and motors may be defined by:

$$\eta_t = \frac{\text{Power out}}{\text{Power in}}$$

A simple relationship between the above-defined efficiencies can be derived. The theoretical work per radian carried out in a rotating pump equals the theoretical torque on the pump input shaft:

$$T_P = \frac{D_P}{2\pi} \cdot p_P$$

The theoretical, required input power E_P on the pump input shaft is expressed by:

$$E_P = \omega_P \cdot T_P$$

Analogously, the effective (required) power E_{eP} on the pump input shaft is:

$$E_{eP} = \omega_P \cdot T_{eP}$$

The theoretical, averaged pump delivery per time unit Q_P is derived by:

$$Q_P = \frac{\omega_P}{2\pi} \cdot D_P$$

The theoretical power E_V transferred to the hydraulic fluid in the pump is:

$$E_V = E_P = Q_P \cdot p_P$$

Analogously, the effective power E_{eV} transferred to the hydraulic fluid in the pump is:

$$E_{eV} = Q_{eP} \cdot p_P$$

Then the total efficiency for a pump can be expressed by:

$$\eta_{tP} = \frac{E_{eV}}{E_{eP}} = \frac{Q_{eP} \cdot p_P}{\omega_P \cdot T_{eP}}$$

Furthermore, by definitions the following relations are derived:

$$T_{eP} = \frac{T_P}{\eta_{mP}} = \frac{\dfrac{D_P}{2\pi} \cdot p_P}{\eta_{mP}}$$

$$Q_{eP} = \eta_{vP} \cdot Q_P = \eta_{vP} \cdot \omega_P \cdot \frac{D_P}{2\pi}$$

Finally the total degree of efficiency η_{tP} for a pump is now expressed as the product of the volumetric degree of efficiency η_{vP} and the mechanical degree of efficiency η_{mP}:

$$\eta_{tP} = \frac{Q_{eP} \cdot p_P}{\omega_P \cdot T_{eP}} = \frac{\eta_{vP} \cdot \omega_P \cdot \dfrac{D_P}{2\pi} \cdot p_P}{\omega_P \cdot \dfrac{\dfrac{D_P}{2\pi} \cdot p_P}{\eta_{mP}}} = \eta_{vP} \cdot \eta_{mP}$$

Analogously, this relationship also hods for a motor:

$$\eta_{tM} = \eta_{vM} \cdot \eta_{mM}$$

5.3 Practical performance characteristics for pumps and motors

For many years researchers have tried to formulate analytic, stationary torque, flow and power loss models for hydraulic motors and pumps. As in many other engineering topics the lack of consistency in formulating mathematical models for the performance of hydraulic machines has concentrated research efforts on experimental identification of such performance characteristics. The primary concept of these is via measurements to illustrate in a quantitative way how the losses of a specific pump (or a motor) will vary with the operational conditions.

In Fig. 5.3 a diagram of performance characteristics for a hydraulic, positive pump is shown. The characteristics are curves of constant, effective pump output flow rate, Q_{eP}, and of constant pump output hydraulic pressure, p_P. The curves are plotted in a normal orthogonal coordinate system with pump input speed, n_P, as the abscissa and effective pump torque, T_{eP}, as the ordinate. In Fig. 5.3a the characteristics for a loss-free (ideal) pump are shown. In this case the curves become horizontal and vertical straight lines. In Fig. 5.3b the characteristics for a real pump are shown in principle. Now the curves deviate from being straight lines due to volumetric and mechanical losses.

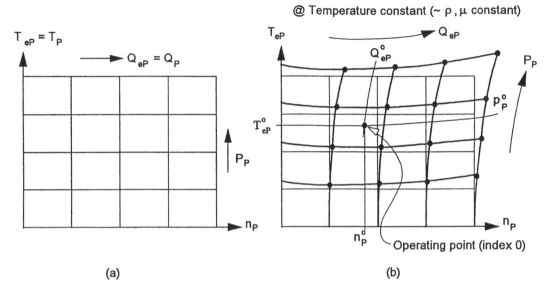

Fig. 5.3 Stationary performance characteristics for a pump

Assuming that the curves in Fig. 5.3b depict the measured characteristics of a real, specific pump (displacement D_P) the efficiencies can be derived from the diagram. Suppose the operating point is as indicated; then the operating values, Q_{eP}^0, p_P^0, n_P^0, T_{eP}^0, can be identified from the diagram and the efficiencies may be computed by using the definitions in section 5.2:

$$\eta_{tP} = \frac{Q_{eP}^0 \cdot p_P^0}{T_{eP}^0 \cdot 2\pi \cdot n_P^0}$$

$$\eta_{mP} = \frac{p_P^0 \cdot \dfrac{D_P}{2\pi}}{T_{eP}^0}$$

$$\eta_{vP} = \frac{Q_{eP}^0}{2\pi \cdot n_P^0 \cdot \dfrac{D_P}{2\pi}} = \frac{Q_{eP}^0}{D_P \cdot n_P^0}$$

The stationary evaluation of hydraulic machines using performance characteristics as shown in Fig. 5.3 requires relatively accurate measurements of the four main variables flow, torque, speed and pressure under steady-state operational conditions for the measured object. This means that the viscosity, μ, and the density, ρ, of the pressure medium must be kept constant during the measurements.

In practice these conditions are in most cases met by keeping the temperature of the pressure medium constant.

A practical example of performance curves for a radial piston pump is shown in Fig. 5.4. Note that in such a diagram, curves for constant input power, E_{inP}, to the pump and curves for constant efficiencies are superimposed as shown.

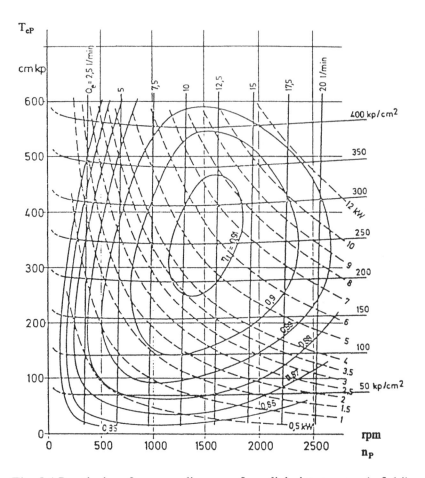

Fig. 5.4 Practical performance diagram of a radial piston pump (ref. 14)

Diagrams of this type are very helpful to the user for identifying how the degree of efficiency varies in the operational range in which he or she plans to use the pump.

5.4 Water hydraulic pumps

Classical three-throw piston water pumps

Hydraulic presses were known in the early development of production machine design. The pressure medium available was water. Today heavy hot-forming presses still often operate with water, mainly to avoid fire hazards. The water pumps originally used for hydraulic presses were made for power ratings in the range of 50 kW and up to approximately 1000 kW. Typically a pump supplies an accumulator water bottle from which pressure water is drawn by the press at the required rate. The pressures may be as high as 300–400 bar.

The classical design of water pumps for heavy hot-forming press was often based on a three-throw piston pump with a horizontal layout. The pistons are driven via a crankshaft mechanism. The prime mover may be an AC electric motor which, via a gear, drives a crankshaft at a speed typically of 120–180 rpm.

1 - inlet valves
2 - cylinder seal
3 - piston (plunger)
4 - cylindrical slideway
5 - crosshead
6 - bearing shell
7 - connecting rod
8 - crankshaft
9 - screw sleeve
10 - oil drain
11 - hemispherical bush
12 - retaining flange
13 - distance piece
14 - water receiver
15 - check valve

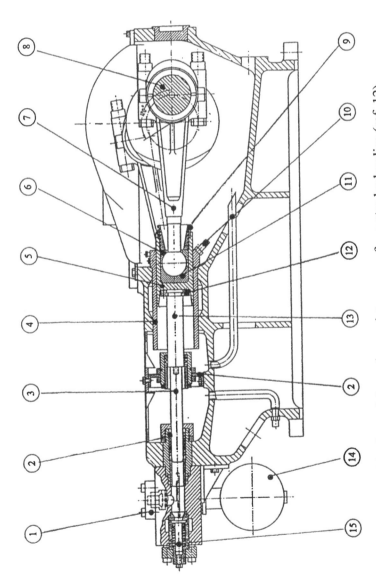

Fig. 5.5 Classical horizontal three-throw piston pump for water hydraulics (ref. 12)

A schematic cross-section of such a slow rotating pump is shown in Fig. 5.5. As can be seen, the mechanical drive including the oil lubrication to the bearings and the crosshead slide, is separated from the pump itself. In this way the water in the water-based hydraulic system is completely separated from the lubrication oil in the crankshaft mechanism. The crankshaft has the three positions displaced 120 degrees. This gives the pump design a fairly building length and a reasonably smooth flow delivery rate.

The plunger material is either flame-hardened mild steel or nitride-hardened Nitraloy steel with a hardness up to 500–600 Brinell. The packings used are multi-vee or chevron packings made of chrome leather or of molded synthetic rubber such as Sumrit.

Under optimal operating conditions three-position piston pumps may have a total efficiency as high as 0.97. When regularly maintained, pumps of this classical design have extremely long lifetimes (several decades).

Modern water hydraulic pumps

The increasing interest over the last decade in using fluid power systems with plain water as pressure medium has challenged manufacturers to develop, design, produce and market a range of pumps, motors and control valves operating entirely on plain water.

In this process two different directions have been taken. One direction been based on using designs originally developed for applying mineral oil as the pressure medium and then gradually introducing modifications in the designs so that the employed pressure medium can contain an increasing amount of water. But in this way the full benefit of using plain water (tap water) may never be reached.

The other direction is based upon introducing radical changes in the component designs by using new materials, new design principles and construction details in order to achieve optimal performance by using plain water as the pressure medium.

Pumps and motors provide the exchange from mechanical torque to fluid pressure and vice versa. They are prone to leakages due to the low viscosity of water caused by high pressure differences across input and output ports. It is therefore important to ensure that leakage resistance between high-pressure ports and low-pressure ports is maximized, i.e., that passages where leakage might occur must be long and that leakage gaps must be kept small.

The above requirement may be more easily met using piston/cylinder mechanisms in water hydraulic pumps and motors rather than gear and vane mechanisms. This probably explains why piston pumps and motors seem to be the preferred designs for such machines.

The following three types of water hydraulic pumps on the market are described:

(a) Crankshaft-driven, triple-throw piston pumps
(b) Radial piston pumps
(c) Axial piston pumps

(a) Hauhinco triplex-piston pumps

Pumps based upon the classical piston pump design with a crankshaft are still used in water hydraulic systems and are competitive on the market. The design has been proven over many years of use and the construction has continuously been updated. A modern type of piston pump, a triplex-piston pump, EMP-3K, manufactured by Hauhinco Maschinenfabrik (Germany), is illustrated in Fig. 5.6. The pump contains 3 pistons arranged in a horizontal layout.

Fig. 5.6 Hauhinco triplex-piston pump

The pumping movements of the pistons are generated by a three-throw, short-stroked crankshaft by an electromotor via a mechanical reduction gear (see Fig. 5.7).

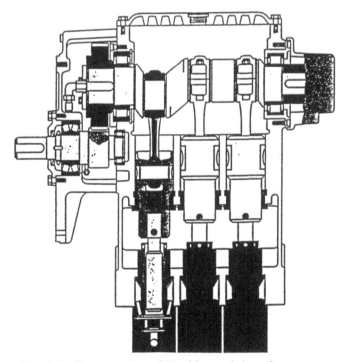

Fig. 5.7 Cross-section of Hauhinco triplex-piston pump

The lubrication system for the crankshaft mechanism and the gearbox is mechanically segregated from the water pump mechanism, so oil cannot enter the water hydraulic system.

A detailed assembly of a cylinder unit is as a cross-sectional view shown in Fig. 5.8.

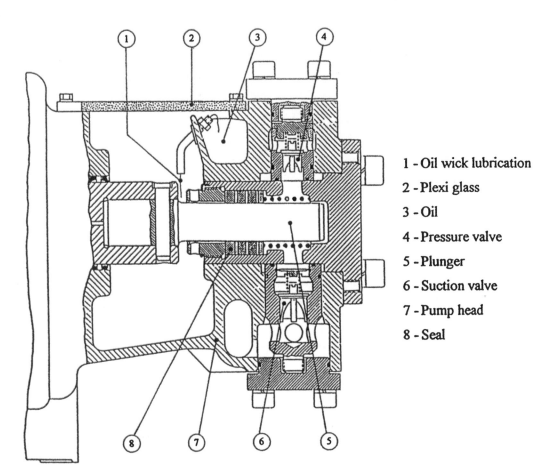

1 - Oil wick lubrication

2 - Plexi glass

3 - Oil

4 - Pressure valve

5 - Plunger

6 - Suction valve

7 - Pump head

8 - Seal

Fig. 5.8 Assembly of a cylinder unit (Hauhinco)

The in- and outflow of fluid is controlled by seat valves of the butterfly type. A Plexiglas window allows visual inspection of the pump seal assembly and the condition of the piston. Note the oil wick lubrication of the piston seal washer. All the vital components are made of corrosion-resistant materials.

The Hauhinco triplex-piston pumps can operate with the following pressure media: HFA fluids, raw water and all neutral fluids in the viscosity range 0.5 to 4 cSt (4 mm^2/sec).

Typically pump ratings are:

- Flow rate up to 700 l/min
- Pressure rate up to 800 bar
- Power up to 200 kW

Typical application areas of this pump type are:

- Mining
- Steel industry
- Automotive industry
- Rubber, wood and textile industry

(b) Hauhinco Radial Piston Pump

A radial piston pump product line, RKP-40 to RKP-160, for operation with raw water has been developed and marketed by Hauhinco Maschinenfabrik (Germany). The pump design (patent pending) is based upon using new and non-traditional materials so that a separate lubrication system is avoided. The medium, industrial water, to be pumped is also the lubricant.

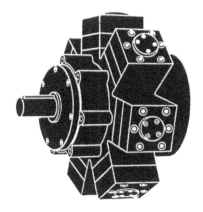

Fig. 5.9 Hauhinco radial piston pump (patent pending)

The RKP pumps are self-priming check valve–controlled radial piston pumps with 5 or 7 piston units. The pump (Fig. 5.9) consists principally of a body in which 5 (or 7) equally spaced cylinders are radially positioned. In Fig. 5.10 a cross-sectional view of the pump is shown. The cylinders surround the eccentric portion of the driving shaft. When the shaft rotates, the eccentric generates the pumping movement of the pistons. Each cylinder houses an accurately fitted piston (plunger), which at its inner end is spherical formed and is held by a spring and a retaining ring against a sliding shoe on the outer surface of the eccentric. The pressure and suction valves are designed as plate valves in order to minimize the dead volume around the cylinders.

The Hauhinco radial piston pumps have a fixed displacement, but by adjusting the length of the piston many different flow rates can be achieved. The pumps are directly driven by electric motors without using gears.

The Hauhinco radial piston pumps can be applied in hydraulic systems using raw water.

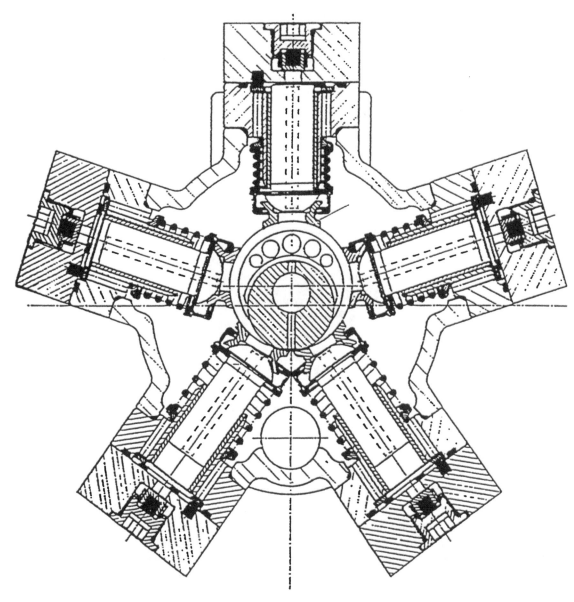

Fig. 5.10 Cross-section of Hauhinco radial piston pump

In Fig. 5.11 a static performance diagram of a Hauhinco radial piston pump is shown. Typically pump ratings are:

- Flow rate up to 242 l/min
- Pressure up to 320 bar
- Power up to 110 kW

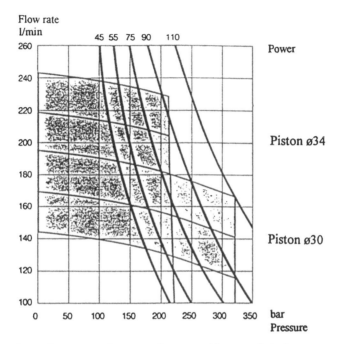

Fig. 5.11 Static performance diagrams for a Hauhinco radial piston pump, RPK-160

(c1) Nessie® Axial Piston Pump

The Danfoss Nessie pump of type PAH (pump axial high-pressure) is designed as an inline axial unit based upon the swash plate principle (see Fig. 5.12). The term **inline pump** means that the center-line axis of the pistons and cylinders is parallel to the center-line axis of the input drive shaft of the pump. The pump is designed especially for using **plain water (tap water) as the hydraulic fluid**. The pump may be used in separate hydraulic systems where the water, after having being used as pressure medium in a motor or cylinder, recirculates to a reservoir and is sucked into the pump again. The pump may just as well be used in hydraulic systems where fresh water is continuously being fed into the system and harmlessly piped to the drain after use.

The present type of pump is with fixed displacement and contains 9 pistons, whereby the kinematic degree of irregularity in the displacement has been minimized to 1.5%. In the following description refer to Figs. 5.12 and 5.13.

The pump input shaft and cylinder barrel are integrated into one piece and made of stainless steel. The cylinder barrel is mounted in a bearing in the outer housing. In order to compensate for slants and deformations from load forces on the cylinder barrel, a loose thrust plate, which rotates with the barrel, is held against the stationary valve port plate via a spring force. This plate is made with the characteristic kidney-shaped openings, which provides for the timely correct commutation when the axial pistons displace hydraulic fluid flow into the pump outlet port during the rotation of the barrel.

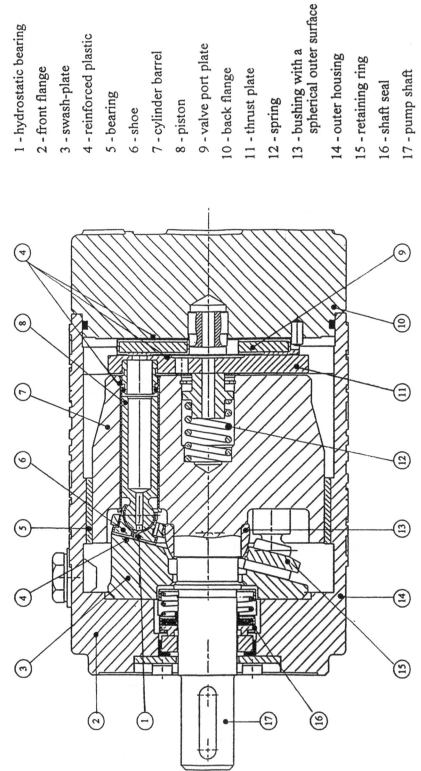

1 - hydrostatic bearing
2 - front flange
3 - swash-plate
4 - reinforced plastic
5 - bearing
6 - shoe
7 - cylinder barrel
8 - piston
9 - valve port plate
10 - back flange
11 - thrust plate
12 - spring
13 - bushing with a spherical outer surface
14 - outer housing
15 - retaining ring
16 - shaft seal
17 - pump shaft

Fig. 5.12 Danfoss's Nessie® axial piston pump (patent pending). Cross-sectional view

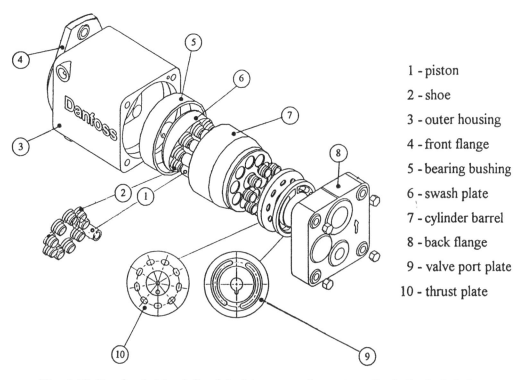

1 - piston

2 - shoe

3 - outer housing

4 - front flange

5 - bearing bushing

6 - swash plate

7 - cylinder barrel

8 - back flange

9 - valve port plate

10 - thrust plate

Fig. 5.13 Danfoss's Nessie® axial piston pump (patent pending). Exploded view

The swash plate is stationary but replaceable so that different displacements may be achieved. The swash plate angle with the pump shaft center line determines the length of the piston stroke. When the cylinder barrel rotates, the shoes mounted on the ball-shaped piston ends slide on the swash plate. Due to the spring force between the thrust plate (a pressure- compensating plate) and the cylinder barrel, the bushing with a spherical outer surface mounted on the barrel and a retaining ring will hold the piston shoes tight to the swash plate. The shoes are provided with hydrostatic bearings.

All friction surfaces, such as cylinders, thrust plate and piston shoes, are made solely of reinforced plastic or overmolded stainless steel.

The outer housing and end flanges are made of chronite-casted special brass in order to prevent corrosion. The pump shaft seal is of a standard mechanical type.

The Nessie axial piston pump is uniquely designed for supplying water flow under high pressure in water hydraulic systems. The pump is designed so that the lubrication of the moving parts in the pump is maintained by the water itself. By the selection of the right materials, the gaps between the moving parts are kept very small in order to minimize leakage losses (because of the extreme low viscosity of water in comparison with mineral oil). On the other hand, the materials selected are able to sustain the generation of hydrodynamic and static fluid films between the moving parts even under high loads and are able to exhibit acceptably low dry friction and rate of wear so that mechanical losses are minimized.

All the parts in the pump are made of non-corrosive materials, ensuring a long life of the pump.

In Fig. 5.14 the Nessie® pump line is shown, and in Table 5.1 a set of technical specifications for these pumps is compiled.

Fig. 5.14 The Nessie® pump line

Table 5.1

Technical specifications for Nessie® pumps

PUMP type	10	12.5	25	32	63	80
Displacement (cm³/rev)	10	12.5	25.3	32.5	63.3	80.4
Max. pressure. cont. (bar)	160	160	160	160	160	160
Max. speed. cont. (rpm)	1500	1500	1500	1500	1500	1500
Min. speed. cont. (rpm)	700	700	500	500	500	500
Max. flow. cont. (l/min) @ 160 bar	13.7	17	34	44	86	112
Input power requirement (kW) @ 1500 rpm/160 bar	4.2	5.3	10.5	13	26	33
Min. suction pressure. cont. (bar abs.)	0.9	0.9	0.9	0.9	0.9	0.9
Max. suction pressure. cont. (bar abs.)	7	7	7	7	7	7
Generated flow ripple (%)	5 (5 pistons)	5 (5 pistons)	1.5 (9 pistons)	1.5 (9 pistons)	1.5 (9 pistons)	1.5 (9 pistons)
Weight (kp)	4.8	4.8	18	18	37	37
Length incl. shaft (mm)	200	200	248	248	320	320
Width. house square (mm)	105	105	128	128	160	160

In order to illustrate the application ranges for the Nessie pump, typical performance data for the pumps are given in Figs. 5.15 and. 5.16.

In Fig. 5.15 (a) and (b) an overview of the delivery rates and mechanical input power requirements for the Nessie pump line from 0-bar and 160-bar pressure load, respectively is given.

As an example, in Fig. 5.16 (a) and (b) the total efficiency, η_{tP}, and the volumetric efficiency, η_{vP}, for a Nessie pump (type PAH 32) are shown, respectively, as a function of pressure and speed.

In recent years regulatory authorities in many countries have begun specifying maximum noise levels for machinery in many application areas. In water hydraulics systems and components it should be noted that the low viscosity of water makes such systems inherently noisier than oil hydraulic systems, and therefore special care must be addressed at the design stage to dampen sources of noise.

The Nessie pumps have been tested at a sound-isolated function panel at the manufacturer's site. A typical suite of test data is shown in Fig. 5.17, where noise sound pressure levels have been recorded for two sizes of pumps @ 100- and 160-bar pressure loads across a speed range of 1000 to 1800 rpm. The maximum sound pressure level was recorded at 79 dB(A).

Lifetime expectations are better than 8000 operating hours @ 160-bar pressure load and at a speed range of 1500–1800 rpm.

(c2) Fenner axial piston pump

The Fenner company in the United Kingdom (see Appendix E) has developed an axial pump for raw water with fixed displacement and either 9 or 5 cylinders, depending on displacement. A range of four sizes of pumps (and motors) exists (Table 5.2). Pumps and motors are virtually identical.

Table 5.2

Technical data for Fenner pumps (and motors)

Characteristic parameter	Unit	PUMPS/MOTORS			
		F06	F15	F30	F60
Displacement	cm³/rev	6	15	29	63
Speed	rpm	1500	1500	1500	1500
Delivery/supply	l/min	9	22	40/48	86/103
Pressure	bar	140	140	140	140
Torque	Nm	13	33	69/62	149/131
Diameter	cm	6.5	13	16	21
Length	cm	11	15	19	24
Weight	kp	4.5	11	22	44
Power	kW	2	5	11/10	23/20

The pumps (and motors) are of the swash plate type. Bearing centers are designed to be in line with the torque centers of the units in order to minimize tilt forces on the valve plate. All rubbing surfaces use polymer–stainless steel interfaces. An adaptation of the pump F30 using ceramic pistons has operated successfully at up to 140 bar. The casing are made of stainless steel. For critical weight applications aluminum and plastic may be available.

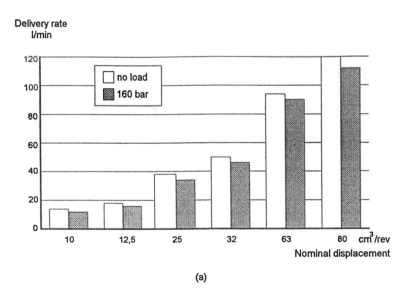

(a)

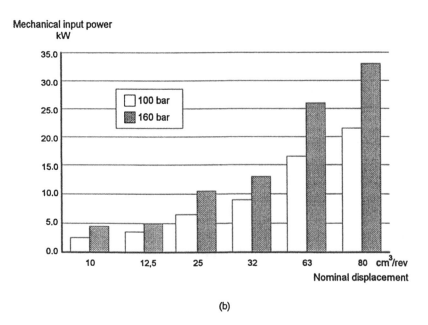

(b)

Fig. 5.15 (a) Delivery rates for Nessie pump type PAH 10–18 cm³/rev.
@ 1500 rpm and 40°C
(b) Mechanical input power for Nessie pump type PAH 10–80 cm³/rev.
@ 1500 rpm and 40°C

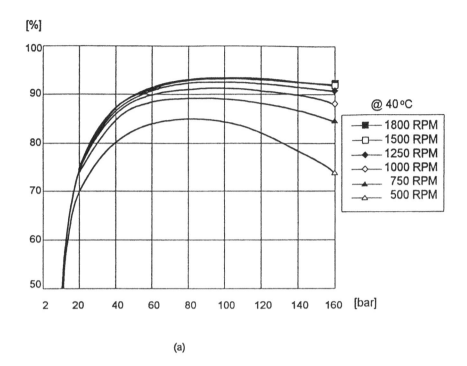

(a)

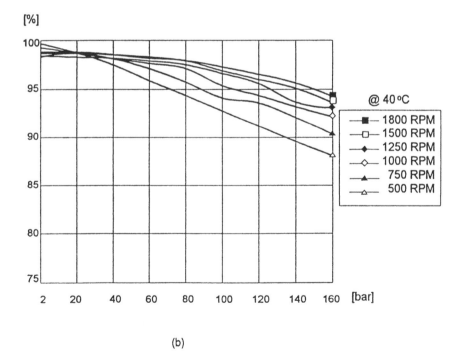

(b)

Fig. 5.16 (a) Total efficiency, η_{tP}, for a Nessie pump (PAH 32) as a function of pressure and speed

(b) Volumetric efficiency, η_{vP} for a Nessie pump (PAH 32) as a function of pressure and speed

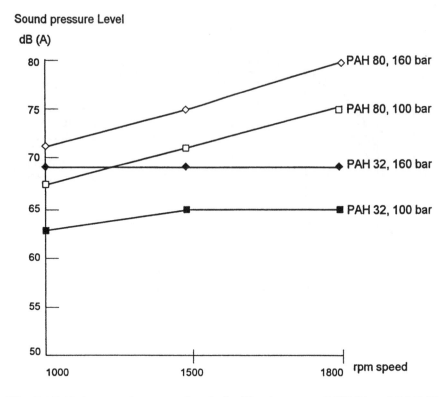

Fig. 5.17 Noise sound pressure levels for Nessie pumps PAH 32 and PAH 80

6. HYDRAULIC MOTORS

In hydraulic systems pumps convert mechanical energy to hydraulic pressure energy and hydraulic actuators convert hydraulic pressure energy back into mechanical energy (see Figs. 1.1 and 1.6).

Assuming that the hydraulic actuators are rotary motors, the hydraulic drive basically transmits mechanical power at **the input state of the pump**, i.e., the product of the speed and the torque of the pump input shaft, into mechanical power at **the output state of the hydraulic motor**, i.e., the product of the speed and the torque of the motor output shaft.

Note that the hydraulic input power to the hydraulic motor is computed as the product of the hydraulic volumetric input flow rate to the motor and the hydraulic pressure.

For practical reasons hydraulic actuators or motors can be divided in two types:

(a) Rotary and semi-rotary motors
(b) Linear hydraulic cylinders

Rotary motors are in principle hydraulic pumps running backwards. In practice the same displacement principles used in pumps are also applicable to rotary motors. Two versions of rotary motors exist: rotary motors where the output shaft carries out continuous angular motions and **semi-rotary motors** where the output shaft is capable of carrying out only limited angular motions such as a fraction of one revolution or just a few of revolutions.

Linear hydraulic cylinders, sometimes called jacks or rams, provide straight-line motions, with the stroke limited by the specific physical construction selected for the cylinder unit in question.

Many of the concepts that apply to pumps apply equally to motors, but there are quite definite differences in their design layouts due to the various operational requirements. Table 6.1 (ref. 9) lists and compares the most important requirements of motors.

The various types of hydraulic rotary motors can be characterized by their different speed ranges. Depending on the application load, the selected hydraulic motor speed range must match the load requirements. Therefore, for a given motor its minimum and maximum speed become important. Hydraulic motors may be categorized in two ways: high-speed/low-torque units and low-speed/high-torque units. In Table 6.2 minimum and maximum speed ranges for these two categories of motors are provided.

6.1 Practical performance characteristics

As noted in section 5.2, the stationary behavior of a hydraulic motor is characterized by four main variables: **torque, speed, volume flow rate** and **pressure**. The variables are evaluated by their average values.

For a rotary motor the ideal stationary relationships are recapitulated. For an ideal motor (subscript M) **the hydraulic power input**, E_{inM}, is expressed by:

$$E_{inM} = p_M \cdot Q_M$$

where p_M is the input hydraulic pressure of the motor (the motor output pressure taken as reference) and Q_M is the average input hydraulic volume flow rate.

Table 6.1

A comparison of requirements of hydraulic motors and pumps

Motors	Pumps
The motor is required to deliver an output torque at a given differential pressure	The pump is required to deliver an output flow at the pressure required by the load
The motor is required to offer a sufficiently high mechanical efficiency over its rated working range	The pump is required to offer a sufficiently high volumetric efficiency over its rated working range
The motor is required to operate over a wide range of output speeds	In an industrial hydraulic system the pump is normally required to operate at a high constant speed
The motor is most often required to meet the highest load pressure at zero or at the low end of its speed range	The pump is required to be able to deliver flow at its highest rated load pressure
The motor is normally required to operate both in clockwise as well as counter-clockwise modes	Except in hydraulic transmissions, the pump is required to rotate in only one direction
The motor may be required to operate in a pumping mode, where the load may overcome pump pressure and reverse motor rotation and motor flow	
Most hydraulic motors may carry only very small side loads on their drive shaft, e.g. from gear wheels or pulleys (Consult the manufacturer!)	Most pumps are permanently directly driven by electric AC motors via elastic couplings. The pump and motor are normally mounted on a common base plate so that side loads on the pump shaft are minimized.
An abrupt startup of a cold motor may be damaging for the motor components	Most pumps normally operate continuously and therefore meet only slowly changing temperatures
Required capacity for motors: Motor displacement D_M	Required capacity for pumps: Pump flow and speed $(D_P = Q_P / n_P)$

Table 6.2

Overview of characteristic ranges for minimum and maximum speeds of hydraulic motors

	Speed ranges for rotary motors	
Motor category	Minimum speed range rpm	Maximum speed range rpm
High-speed/low-torque unit (e.g., swash plate piston motor)	$200 \rightarrow 500$	$3000 \rightarrow 8000$
Low-speed/high-torque unit (e.g., radial piston motor)	$0.1 \rightarrow 10$	$200 \rightarrow 1000$

The mechanical power output of an ideal motor, E_{outM}, is expressed by:

$$E_{outM} = T_M \cdot \omega_M$$

where T_M is the output torque on the motor shaft and ω_M is the angular velocity (in radians per unit time).

The relationship between the motor displacement D_M in volume units per motor shaft revolution and the angular speed, ω_M, of the motor shaft is by definition:

$$D_M = \frac{Q_M \cdot 2\pi}{\omega_M}$$

The theoretical output torque on the motor shaft is derived by:

$$T_M = \frac{D_M}{2\pi} \, p_M$$

Practical performance characteristics for motors are based on experimental identification rather than on analytic mathematical models for flow and torque losses. As noted for pumps in section 5.3, this is due to lack of consistency in the mathematical models that are available.

In Fig. 6.1 a diagram of performance characteristics for a hydraulic, positive, rotary motor is shown. The characteristics are curves of constant, effective motor input flow rate, Q_{eM}, and of constant motor input hydraulic pressure, p_M^0. The curves are plotted in a normal orthogonal coordinate system, with motor output speed, n_M, as the abscissa and effective motor output torque, T_{eM}, as the ordinate. In Fig. 6.1a the characteristics for a loss-free (ideal) motor is shown. In this case the curves become horizontal and vertical straight lines. In Fig. 6.1b the characteristics for a real motor are shown in principle. Now the curves deviate from being straight lines due to volumetric and mechanical losses.

Assuming that the curves in Fig. 6.1b depict the measured characteristics of a real, specific motor with the displacement D_M, the motor efficiencies can be derived from the diagram. Suppose the operating point for the motor is as indicated. Then the operating values Q_{eM}^0, p_M^0, n_M^0, and T_{eM}^0 can be identified from the diagram. The motor efficiencies may be computed by using the definitions in section 5.2:

$$\eta_{tM} = \frac{T_{eM}^0 \cdot 2\pi \cdot n_M^0}{Q_{eM}^0 \cdot p_M^0}$$

$$\eta_{tM} = \frac{T_{eM}^0}{p_M^0 \, p_M^0}$$

$$\eta_{vM} = \frac{2\pi \cdot n_M^0 \cdot \dfrac{D_M}{2\pi}}{Q_{eM}^0} = \frac{D_M \cdot n_M^0}{Q_{eM}^0}$$

Note that:

$$\eta_{nM} \cdot \eta_{vM} = \eta_{tM}$$

The stationary evaluation of hydraulic motors using performance characteristics as shown in Fig. 6.1 requires relatively accurate measurements of the four main variables flow rate, torque, speed and pressure under steady-state operational conditions for the measured object, i.e., hydraulic fluid viscosity and density must be kept constant.

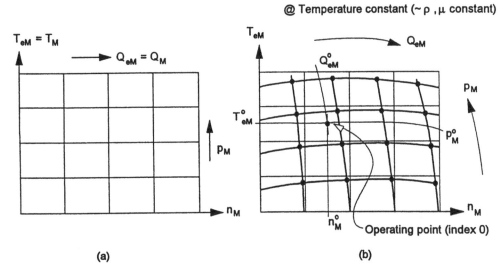

(a)   (b)

Fig. 6.1 Stationary performance characteristics for a hydraulic motor (displacement D_M)

A practical example of performance curves for a hydraulic motor is shown in Fig. 6.2. In this diagram the curves for constant mechanical output power, E_{outM}, from the motor and the curves for constant total efficiencies, η_{tM}, are superimposed as shown. The curves for constant efficiencies often are closed loops, as in Fig. 6.2. Because of its shape, it is called a **mussel diagram**.

A diagram of this type is very helpful to the user because it helps to overview the efficiency variations for the hydraulic motor over the full application range.

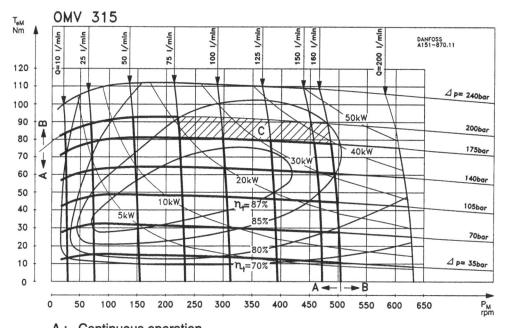

Fig. 6.2 An example of measured performance characteristics for an oil hydraulic motor (mussel diagram)

6.2 A water hydraulic motor

Several research projects for developing hydraulic motors that operate on water (tap water, industrial water) have been initiated worldwide. As of spring 1995, only one, the Danfoss Nessie® swash plate axial piston motor, type MAH (motor axial high-pressure), has arrived on the commercial market. The axial unit is a so-called in-line motor, which means that the center line axis of the pistons and cylinders are parallel to the center line axis of the output shaft of the motor. The motor is designed especially for using **tap water as the hydraulic fluid**.

The present type of motor is with fixed displacement and contains 5 pistons. The kinematic degree of irregularity in the displacement as a function of motor shaft rotation is expressed as the ratio of the ripple amplitude over the mean value of the generated flow: 4.96 %.

In Fig. 6.3 is shown a cross-sectional view of the Danfoss Nessie® swash plate axial piston motor, type MAH. Basically the motor design is very much the same as the Nessie® pump shown in Fig. 5.12.

The motor output shaft and the cylinder barrel are integrated into one piece and are made of stainless steel. The cylinder barrel is mounted on two hydrostatic journal bearings, one bearing at the end of the housing and another bearing in the port flange. In order to compensate for slants and deformations from pressure load forces on the barrel, a loose thrust plate rotating with the barrel is held against the stationary port plate via a spring force. This plate is made with the characteristic kidney-shaped openings which provide the timely correct commutation so that the axial pistons are displaced by the hydraulic pressure fluid entering the motor inlet port and cause the barrel to rotate.

The swash plate is stationary but replaceable so that different displacements may be achieved. The swash plate angle with the motor shaft center line determines the length of the piston stroke. When the cylinder barrel rotates, the shoes mounted on the ball-shaped piston end slide on the swash plate. Due to the spring force between the thrust plate (a pressure-compensating plate) and the cylinder barrel, the bushing with a spherical outer surface mounted on the barrel and retaining ring will hold the piston shoes tight to the swash plate. The shoes are provided with hydrostatic bearings.

All friction surfaces, such as cylinders, thrust plate and piston shoes, are made solely of reinforced plastic solely or overmolded stainless steel.

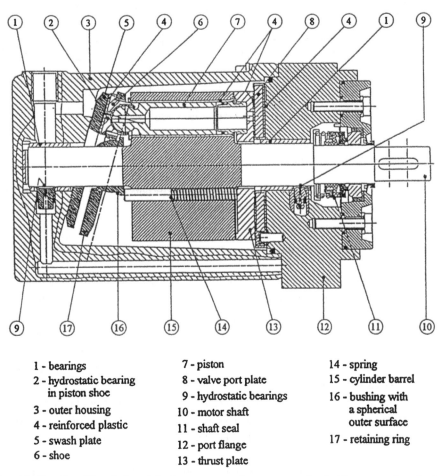

1 - bearings	7 - piston	14 - spring
2 - hydrostatic bearing in piston shoe	8 - valve port plate	15 - cylinder barrel
3 - outer housing	9 - hydrostatic bearings	16 - bushing with a spherical outer surface
4 - reinforced plastic	10 - motor shaft	
5 - swash plate	11 - shaft seal	17 - retaining ring
6 - shoe	12 - port flange	
	13 - thrust plate	

Fig. 6.3 Danfoss Nessie® axial piston motor (patent pending). Cross-sectional view

Housing and flanges are made of sand-casted aluminum, which is trowelled and pulse-anodized for an optimal corrosion protection. The motor shaft seal is of a standard mechanical type.

The motor is designed so that the lubrication of the moving parts in the motor is maintained by the water itself. By the selection of the right materials, the gaps between the moving parts are kept very small in order to minimize leakage losses (because of the extremely low viscosity of water in comparison with mineral oil). On the other hand, the materials selected are able to sustain the generation of hydrodynamic and static fluid films between the moving parts even under high loads and are able to exhibit acceptably low dry friction and rate of wear so that mechanical losses are minimized.

In Table 6.3 technical data for Nessie® hydraulic motors, type MAH, are presented and in Fig. 6.4 an engineering drawing of the front and side views of a Nessie® hydraulic axial piston motor, type MAH 10/12.5 is shown. This type of motor is of very compact design and of low weight. The motor is fully lubricated by the hydraulic water flow and service life comparable with that of traditional oil hydraulic motors on the market.

Table 6.3

Technical data for Nessie® motors type MAH

MAH type		4	5	6.3	8	10	12.5
Geometric displacement (cm³/rev)		4	5	6.3	8	10	12.5
Max. speed (rpm)	cont.	3500	3500	3500	3000	3000	3000
Max. torque (Nm)	cont.	8.3	10.5	13.3	17	21	25.5
Max. power (kW)	cont.	3	3.8	4.9	5.3	6.4	7.8
Max. pressure drop (bar)	cont.	140	140	140	140	140	140
Max. water flow (l/min)	cont.	17.5	21	25.5	27.5	33	41
Starting torque (Nm) @ max. press. drop	cont.	4	6	8.5	8.5	12	16.5
Min. speed (rpm)		300	300	300	300	300	300
Weight (N)		27	27	27	40.7	40.7	40.7
Filtration rating		\multicolumn{6}{c}{$10 \mu (\sim\beta_{10} = 75)$}					

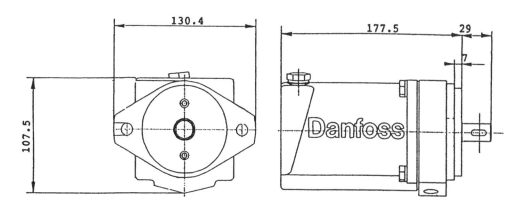

Fig. 6.4 Front and side views of a Danfoss Nessie® axial piston motor, type MAH 10/12.5 (patent pending). Dimensions in mm

This motor is made from non-corrosive materials and has smooth surfaces in order to meet hygienic requirements for external cleaning. It is designed to operate uni-directionally in its primary mode, but it can run in both directions. In Fig. 6.5 a diagram illustrates the efficiencies for a Nessie hydraulic motor, type MAH 12.5 running in the primary direction of rotation at 140 bar. The motor can run in the secondary direction of rotation under light duty cycles at 140 bar up to 1000 rpm.

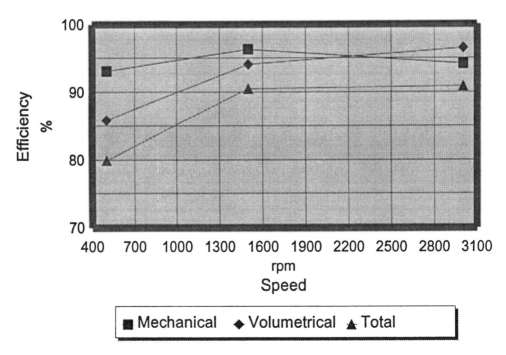

Fig. 6.5 Diagram of efficiencies for a Nessie® motor, type MAH 12.5, at 140 bar

In Fig. 6.6 an overview of the performance characteristics for a Nessie motor, type MAH 12.5, is shown as a mussel diagram. The diagram is basically an orthogonal co-ordinate system with motor speed n_M as abscissa and motor output torque T_{eM} as ordinate. The diagram further depicts lines (curves) for effective motor input flow rate Q_{eM} (vertical lines) and lines (curves) for constant motor pressure p_M (horizontal lines). Finally the diagram depicts loci of constant total efficiencies η_{tM} and curves for motor output power E_{outM}.

It is well known from other branches of hydraulics that noise emission from components is a major factor of concern. In Fig. 6.7 a diagram of noise levels as a function of speed at different supply pressures is shown for a Nessie hydraulic motor, type MAH 12.5. The noise levels shown all comply well with the requirements of Danish regulatory authorities.

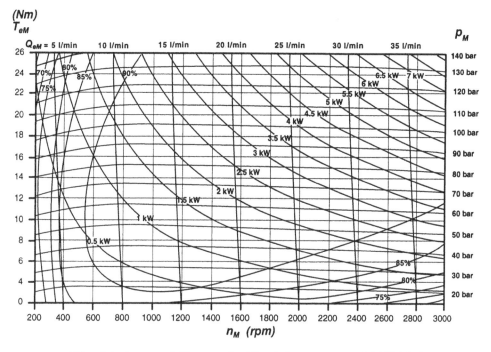

Fig. 6.6 A mussel diagram for Nessie® motor, type MAH 12.5

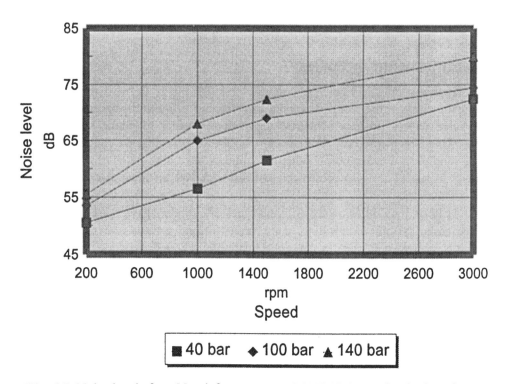

Fig. 6.7 Noise levels for a Nessie® motor, type MAH 12.5, running in the primary direction of rotation

Sometimes the Nessie® hydraulic motor may be better matched to the load by inserting a gear between the load driving shaft and the output shaft of the hydraulic motor. Such an arrangement is shown in Fig. 6.8. The Nessie® gear is a planetary gear specially designed for use in Nessie® water hydraulic systems. The gear is made with corrosion-resistant surfaces. The output shaft is mounted with a double seal, highly resistant to high-pressure cleaning. To ensure problem-free operation in the food industry, the lubricants used are permitted by the U.S. Food and Drug Administration (FDA). Nessie® gears are available for rated starting torque up to 630 Nm and gear ratios from 1:3.51 to 1:38.6.

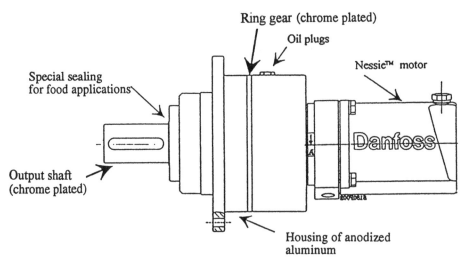

Fig. 6.8 Nessie gear integrated with a Nessie® motor

6.3 Linear actuators, cylinders

Hydraulic cylinders convert hydraulic pressure energy in the hydraulic fluid into mechanical energy in the form of a force carrying out a linear displacement. Hydraulic cylinders are widely used as actuators, especially in water hydraulic systems, mainly because of their high force capabilities and the ease of piston speed control over a wide range of conditions. Furthermore, hydraulic cylinders have high ratios of power to weight and power to size. They also present high stiffness as well as high speed of response to control commands.

As in oil hydraulics, many different types of cylinders have been made for use in water hydraulic systems controlling, e.g., heavy forging presses and presses used for vulcanizing processes. In such systems the hydraulic cylinders are most often integrated parts of the hydraulic press frame itself. Water hydraulic cylinders are also used for various purposes in the mining industry

In more general mechanical automation tasks where water hydraulic systems are used, e.g., in slaughterhouses and escalators, the cylinders are more or less of standard types (see Fig. 6.9).

In principle the same cylinder design that is used in oil hydraulic systems may be used in water hydraulic systems. A cylinder designed for a water hydraulic application can in most cases be applied without changes for oil hydraulics. However, a cylinder designed for oil hydraulics can only in rare cases be used for a water hydraulic application.

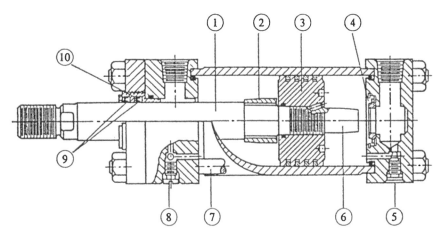

1 - piston rod, case hardened and chrome plated
2 - floating cushion sleeve
3 - piston and piston rings of cast iron
4 - floating cushion bushing
5 - cushion adjusting needle valve
6 - cushion spear
7 - high-tension steel tie rods
8 - cushion check valve
9 - rod seals
10 - removable rod gland follower ring

Fig. 6.9 Typical standard design of a double-acting hydraulic cylinder
with end cushioning (ref. 5)

The main differences in the cylinder design between the two types of applications are due to the special requirements of water hydraulics. Using water as the fluid pressure medium introduces a severe corrosion problem for the components. The materials selected must therefore be of non-corrosive types.

The low viscosity of water compared with the higher viscosity of hydraulic oil requires smaller gaps between moving parts and longer leakage paths in order to minimize leakage in water hydraulic cylinders.

The change of the lubrication conditions between moving parts when using water instead of hydraulic oil requires much smoother surfaces on sliding parts, such as the inner side of the water hydraulic tube, and a different choice of piston sealing material in order to ensure a sufficiently long lifetime of the cylinder.   /

The lower degree of damping for water compared with hydraulic oil poses special requirements for the design of end-stroke cushioning devices with narrow gaps and small tolerances as well as carefully designed throttling valves.

Basic types of cylinders

There are several different types of cylinders. The most frequently found basic types are illustrated in Fig. 6.10. Each type may be available with individual features over a wide range of pressures and with various mounting arrangements.

In Fig. 6.10a a single-acting cylinder with a plunger is shown. Plunger pistons are used in presses and jacks. When the cylinder is mounted in a vertical position with the piston moving upward, the return stroke can be carried out simply by gravitational force. Plunger cylinders often have long strokes and large diameters.

In Fig. 6.10b the cylinder is single-acting and therefore the return stroke must be initiated by external means.

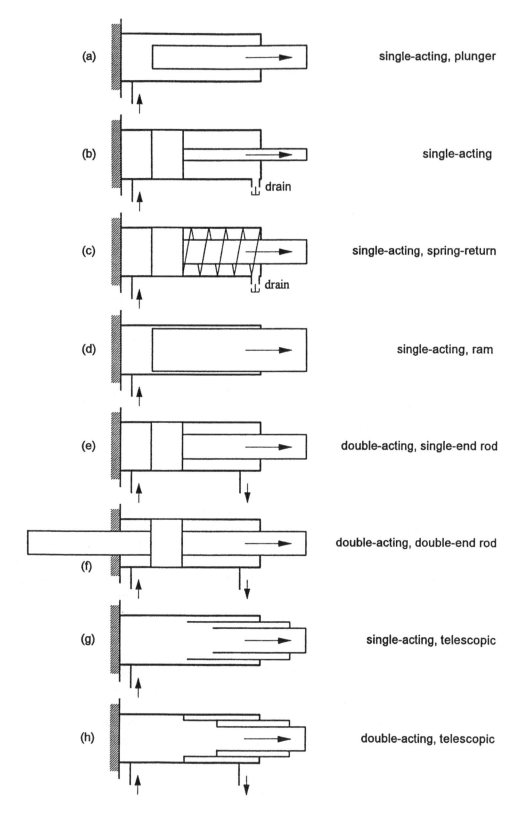

Fig. 6.10 Different types of cylinders

In Fig. 6.10c the cylinder is single-acting and the return stroke is carried by a spring.

In Fig. 6.10d the cylinder is single-acting with the rod diameter equal to the cylinder bore diameter. This type of piston is called a ram. Only one packing, the rod packing, is needed.

In Fig. 6.10e the cylinder is double-acting and the piston can carry out a power stroke in both directions. The piston has a single-end rod so the areas of pressure are uneven. Actuating hydraulic fluid source can be connected to either end of the cylinder, for example, via a 4-port, 2-position valve.

In Fig. 6.10f the cylinder is double-acting and the piston has a double-end rod, which usually offers areas of even pressure. Piston bearings and cylinder seals are required in both ends.

In Fig. 6.10g the cylinder and piston are made of two or more hollow sections. These telescope together when the cylinder and piston are retracted by external means and they extend when the cylinder and piston are hydraulically actuated. The extended length is more than twice that of the retracted length.

In Fig. 6.10h the telescopic unit is double-acting and of the same basic design as the unit in Fig. 6.10g.

Practical considerations

Mounting configurations. One of the big advantages of using standard hydraulic cylinders as actuators in automatic machinery is the flexibility and versatility regarding mounting configurations. Most suppliers of hydraulic cylinders also offer a broad spectrum of classical mounting combinations (see Fig. 6.11).

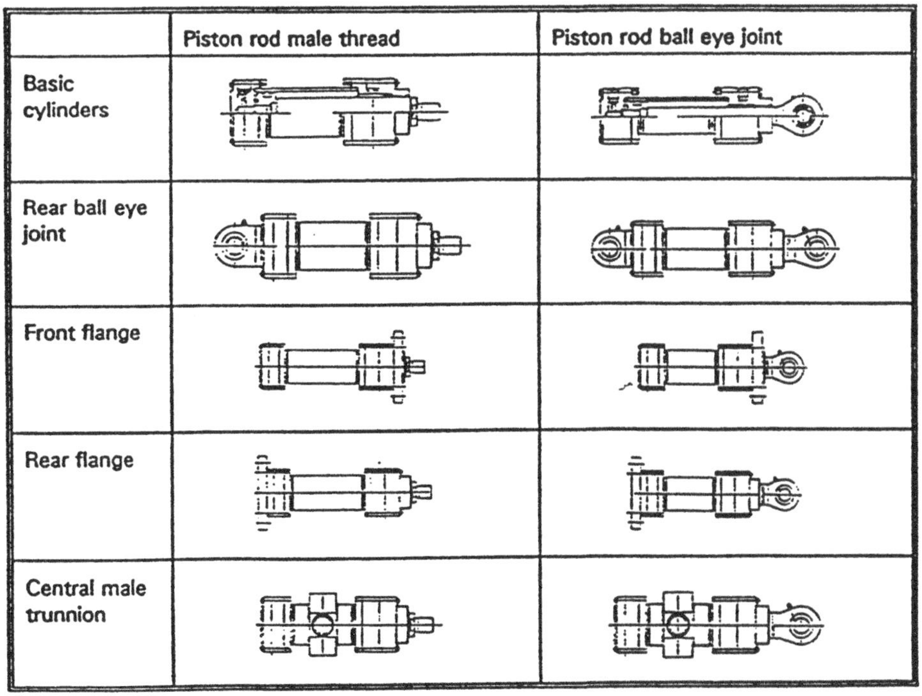

	Piston rod male thread	Piston rod ball eye joint
Basic cylinders		
Rear ball eye joint		
Front flange		
Rear flange		
Central male trunnion		

Fig. 6.11 Various mounting configurations

Nessie® standard water cylinders. Danfoss has marketed a series of water hydraulic cylinders in the Nessie® water hydraulics program. The cylinders are designed for operating with pure tap water as the hydraulic fluid. The cylinders are double-acting, with a single-end rod (see Fig. 6.12), and the design complies with the international standard ISO-6021/1. The cylinders are made with smooth surfaces, which make them extremely easy to clean. The parts of the cylinder in contact with the water are made of corrosion-resistant steel. As an option the Nessie cylinders can also be supplied in all stainless steel.

In Table 6.4 typical technical data for the cylinders are listed.

Fig. 6.12 Nessie® standard water cylinder (Danfoss)

Table 6.4
Typical data for Nessie cylinders

Rated pressure	160 bar
Test pressure	240 bar
Piston diameter range	32–100 mm
Rod diameter range	18–56 mm
Piston speed, max.	0.2 m/sec
Hydraulic fluid	tap water
Temperature range	+1°C to +50°C
Filtration rating	10 μm (~β_{10} = 75)
Mounting orientation	No constraints

Cylinder cushioning. The maximum speed of the piston in a hydraulic cylinder is limited by the maximum rate of flow into and out of the cylinder ports and by the ability of the cylinder caps to absorb the impact forces when the piston motion is stopped by hitting the cylinder caps. In order to limit these impact forces to an acceptable value, experience shows that the piston speed on impact should not exceed 125 mm/sec. When speeds need to be higher, cushioning devices should be inserted in the cylinder circuit so that the kinetic energy of the piston and its load may be absorbed at acceptable forces on the cylinder end caps.

In Fig. 6.13 a simple cushioning principle for a piston motion is shown. The spike on the piston fits with a narrow clearance into the hole in the cylinder end cap. At a certain position of the piston (assumed to move right) the flow out of the cylinder becomes restricted and a back pressure to the right side of the piston is generated. This pressure will brake the motion of the piston and by a correct adjustment of the shown throttle the piston will stop smoothly by landing against the cylinder end cap. Reverse motion of the piston is established by flow through the check valve.

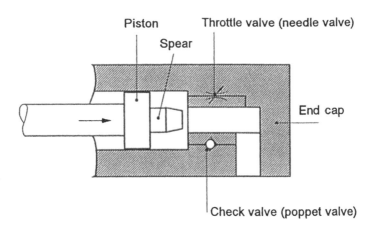

Fig. 6.13 Cylinder cushioning

In water hydraulics the above cushioning principle can be used with good results. The low viscosity of water, however, requires a very small clearance between the hole in the end cap and the piston spike. The throttle valve, most often in the form of a needle valve, must be carefully designed in order to avoid erosion from the water flow. For trouble-free performance of the cushion, the unit may further require a poppet-type check valve.

Water hydraulic cylinder materials. The key to success in water hydraulic technology is the proper selection of suitable types of material for the components. These materials must resist all kinds of corrosion from the water. Furthermore, the parts that slide against each other, frequently under pressure load, must have smooth surfaces that ensure sufficient lubrication films in spite of the rather low viscosity of water, and finally the design must sustain full control of pressure water leakage.

When using pure water, the preferred material for piston, rod, cylinder tube, and cylinder caps is **stainless steel.**

The materials for the various seals are likewise of utmost importance. For static seals, such as O-rings, hard Perbunan materials are frequently used. For dynamic seals such as wiper rings, hard-wearing polyurethane may be preferred. For hydraulic pressures up to 250 bar, Perbunan is used for rod seals of the U-ring type. Piston seals and piston bearing rings are frequently made of PTFE (Teflon).

Parker Hannifin water hydraulic cylinder. hydraulic water cylinder is illustrated in a cross-sectional view in Fig. 6.14. This cylinder contains cushions at both ends of its stroke.

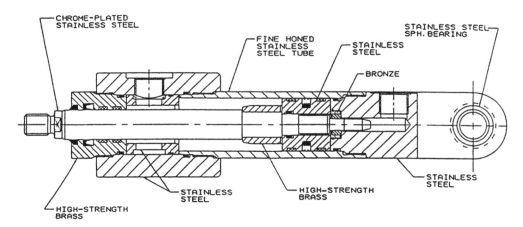

Fig. 6.14 Standard water cylinder with end position cushioning (Parker Hannifin)

7. HYDRAULIC VALVES

The components in a hydraulic system that control the hydraulic power transmission from the pump to the motor (Fig. 9.1) are called valves. Valves may be classified in various ways. One way is to divide valves into categories according to the effect that they execute in the system. Another is to classify valves on basis of their design.

Using the first method, the following three main groups of valves are found:

(1) Directional control valves
(2) Flow-control valves
(3) Pressure-control valves
(4) Proportional control valves

Directional control valves control by their functional positions or states (see, e.g., Fig. 2.4) which motor connections carry pressure flow and return flow. Besides directional flow, such valves also govern other effects in the hydraulic system. The actuation of the valves causes some delays in the control of flow direction due to dead times and finite switching times. Furthermore, directional control valves cause pressure drops and may introduce leakage from the pressure side to the return side.

The function of **flow- and pressure-control valves** is based on restrictions purposely introduced in the hydraulic system. The restrictions may simply be fixed orifices or they can be controllable in size by the pressure, flow, or position of a valve member, such as a spool, poppet, or ball.

The ideal purpose of a flow-control valves is to control the flow in accordance with requirements and thereby the speed of the actuators independently of the pressure.Similarly, the ideal purpose of pressure-control valves is to control the pressure independently of the flow and thereby to protect the hydraulic system against overload.

The ideal pressure-flow ($\Delta p - Q$) characteristics of flow- and pressure-control valves are shown in Fig. 7.1. For pressure-control valves the ideal characteristic is a line parallel with the flow rate axis (a vertical line in Fig. 7.1) and placed at a distance from the origin equal to the adjusted pressure set point (the valve opening pressure). For flow-control valves as well the ideal characteristic is a line parallel with the pressure difference axis (a horizontal line in Fig 7.1) and placed at a distance from the origin equal to the adjusted metered flow rate through the valve (set point).

The ideal purpose of a **proportional control valve** is to deliver either an output flow rate or an output pressure directly proportional to the valve input, which is usually an electronic signal.

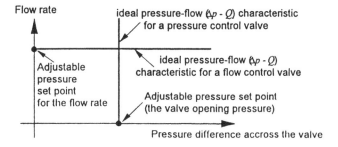

Fig. 7.1 Ideal pressure-flow ($\Delta p - Q$) characteristics of flow and pressure-control valves

In the subsequent sections typical **water hydraulic valves** will be described and their engineering data presented and discussed. Several designs of such valves of different makes have been marketed. With the intention of focussing on the concept of water hydraulics and in order to make this discussion practical, the Nessie® line of hydraulic components is presented as a representative approach to using water hydraulic valves for quite new application areas of controlling machinery by water hydraulics, such as in the food-processing industries.

Water hydraulic valves share many of the basic design principles as oil hydraulic valves. However, using water instead of mineral oil as the pressure medium entails significant changes in the physical parameters, as outlined in Table 7.1.

Table 7.1

Comparison of some physical design parameters for oil and water

	Oil	Water
Viscosity (and lubrication) @ 20°C	~ 30 cSt	~ 1 cSt
Vapor pressure @ 50°C	$1.0 \cdot 10^{-8}$ bar	0.12 bar
Compression modulus @ 20°C	$1–1.6 \cdot 10^{9}$ bar	$2.4 \cdot 10^{9}$ bar
Corrosion protection	Good	Poor

For the design of water hydraulic valves the relatively **low value of viscosity for water** plays a dominant role. This can be seen from the following considerations. Firstly, water's very low viscosity compared with mineral oil will cause a higher flow velocity through a throttle restriction under equivalent conditions. Secondly, the flow of water contains greater kinetic energy (see section 1.3, the Bernoulli equation) compared with mineral oil under equivalent conditions. Thirdly, the leakage through clearances and narrow passages may become excessive in water hydraulic components such as valves. Fourthly, surfaces with relative motion and high perpendicular load pressures may be subject to damaging friction.

The higher energy density of the pressure fluid flow in water hydraulics and the higher vapor pressure of water compared with oil hydraulics may cause serious problems (ref. 1) of **erosion** (via cavitation) and **abrasion** in leakage flows in the functioning of valves. Also the higher energy density may cause **water-hammering problems** with **high transient pressure peaks, noise** and **resonance** in the system.

To limit leakage through a narrow clearance to same amount as in oil hydraulics requires a much closer gap. For comparison consider the circular annulus shown in Fig. 7.2. The leakage can be computed by the **Hagen-Poiseuille formula** assuming laminar flow as follows:

$$Q_{Leakage} = \frac{D \cdot \pi \cdot (\Delta D)^3}{96 \cdot \mu \cdot L} \cdot \Delta p$$

where

$Q_{leakage}$ = leakage flow rate
D = diameter
ΔD = radial clearance
μ = dynamic viscosity
L = length of clearance
Δp = pressure difference across the length of the clearance

Assuming the ratio of viscosities for water, μ_W, and for mineral oil, μ_O

$$\frac{\mu_W}{\mu_O} = \frac{1}{27}$$

The ratio of radial clearances becomes:

$$K = \frac{\Delta D_{Water}}{\Delta D_{Mineral\,oil}} = \frac{1}{\sqrt[3]{27}} \approx 0.33$$

This means that in order to ensure the same amount of leakage using water instead of mineral oil as the pressure medium, a one-third reduction in the radial clearance is required.

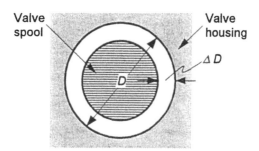

Fig. 7.2 Circular annulus clearance

In general, clearances in water hydraulic components have to be kept very small in order to minimize undesirable leakages and keep efficiencies acceptable. This also necessitates tight tolerances in the manufacturing of the valve elements. In some cases, such as where heat losses from leakage flows might cause temperature deformation of the valve geometry, special material, e.g., ceramic, is used for critical valve elements so that the pressure-flow characteristics of the valve do not degrade when the temperature changes.

To avoid the above problems, water hydraulic valve designs are often of the seating valve type, using, for instance, a ball, poppet, plug, or needle as the valve member. Special attention to built-in damping of the motion of the valve member must also be considered. Note further that the water hydraulic valves are of in-line type and only a few are designed with a subplate for pipe mounting.

7.1 Directional control valves

Basically, directional control valves provide control of when and where fluid power is to be delivered for performing various required functions. They provide pressure fluid flow for starting, stopping, accelerating, and controlling the direction of motion of the hydraulic actuators.

Generally, directional control valves may be classified by the principle of operation and the number of functional positions (see Figs. 2.2–2.4). Various types of control valves may differ considerably in physical characteristics and in operation. Different design principles are used according to the type of the valve member element, such as sliding spool, rotary spool, plate, poppet, and ball.

A specially developed directional control valve for controlling the direction of water flow in water hydraulic systems is shown in Figs. 7.3 and 7.4. This valve is a 2-port, 2-position, pilot-controlled directional valve of a normally closed (NC) version. The pilot stage is with solenoid (DC or AC) actuation and spring return. The main stage is also with spring return. The directional control valve is for use in water hydraulic applications for stop and start of rotary motors and cylinders and for control of their direction of motion. Moreover, this version of valve can be used as a relief valve (unloading valve) in connection with power supplies. Carefully designed, built-in damping ensures a reduction of possible pressure peaks from water flow decelerations.

Fig. 7.3 Nessie® 2-port, 2-position control valve, type VDH 60E

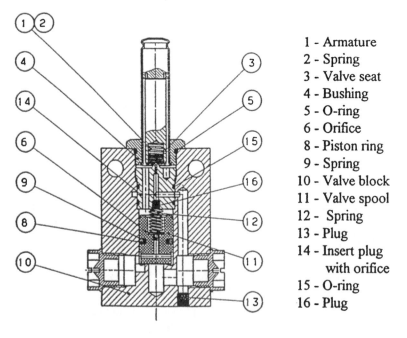

1 - Armature
2 - Spring
3 - Valve seat
4 - Bushing
5 - O-ring
6 - Orifice
8 - Piston ring
9 - Spring
10 - Valve block
11 - Valve spool
12 - Spring
13 - Plug
14 - Insert plug
        with orifice
15 - O-ring
16 - Plug

Fig. 7.4 Cross-sectional view of the valve shown in Fig. 7.3

The valve in Figs. 7.3 and 7.4 is of a seat design with extremely low leakage. The hydraulic fluid to be used is pure water (for water quality specification, refer to EU 80/778) or water added with 33% glycol (DOWCAL N).

The valve block is made of stainless steel ISO 316L and is acid-proof. The block material is in accordance with FDA/3A regulations and is corrosion-proof against the hydraulic fluid and resistant to the cleaning means used within the food industry. The valve block surfaces and the simple form enable simple and effective cleaning. The internal valve details are manufactured of stainless steel, polymers, and brass.

The 2-port, 2-position valve described above may be combined in a block of four valves in a bridge connection, to make a 4-port, 3-position valve. An example of such a bridge connection is illustrated schematically in Fig. 7.5. The connections in the valve's nominal 0-position (see Fig. 7.5b) are all blocked in this state, which is why it is called closed center valve. The 4-port, 3-position directional valve in Fig. 7.5a is pilot operated by two solenoid-actuated pilot valves controlling the four main-stage valves marked with the numbers 1 through 4. Using other combinations of orifices and ball check valves in the four main-stage valves offers other port connections in the nominal 0-position (see Fig. 7.5b).

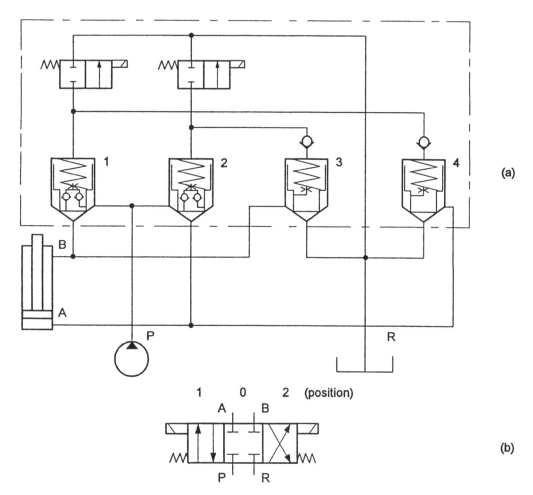

Fig. 7.5 (a) Scheme of four 3-port, 2-position directional control valves bridged together and forming a 4-port, 3-position directional control valve (patent pending)
(b) Standard symbol for the 4-port; 3-position directional control valve

In Fig. 7.6 a cross-sectional view of the physical arrangement of a 4-port, 3-position directional valve is shown.

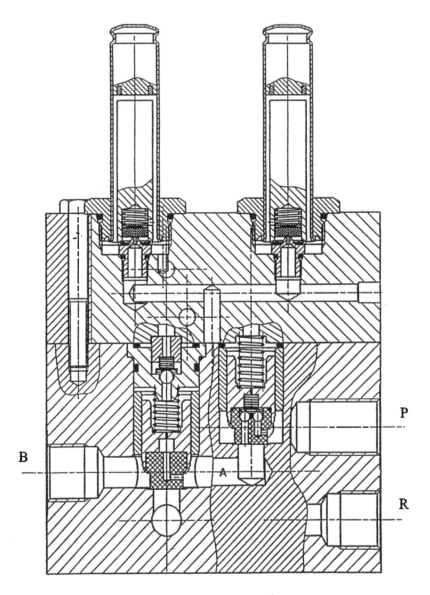

Fig. 7.6 Cross-sectional view of the valve shown in Fig. 7.5 (patent pending)

Pertinent engineering data for 2-port, 2-position directional valves of the type described above and illustrated in Figs. 7.3 and 7.4 are listed in Table 7.2. As can be seen from this table, the performance data for the water hydraulic directional valves are competitive with their oil hydraulic counterparts.

Table 7.2

Technical data for Nessie® 2-port, 2-position directional valves,
type VDH 30E and 60E

Flow capacity	l/min	0–30	0–60
Pressure:			
Max. continuous inlet pressure	bar	140	140
Max. peak pressure	bar	200	200
Opening pressure, max.	bar	1.5	3
Pressure loss @ full flow	bar	4.5	4.5
Leakage, 30 sec @ 140 bar		None	None
Operational mode		ON/OFF	ON/OFF
NC ~ "normally closed"			
or NO ~ "normally open"			
Opening time, max.	msec	70	90
Closing time	msec	200–350	250–400
Electric activation			
Volt DC		12/24	12/24
Volt AC		24/110/220	24/110/220
Lifetime	cycles	$3 \cdot 10^6$	$3 \cdot 10^6$
Temperature range	ºC	2–50	2–50
Level of hygiene		High	High

7.2 Flow-control valves

In water hydraulics flow-control valves are used for controlling the volume flow rate of the water and thereby the speed of actuators. Two types of valves may be used: the pressure-compensated flow-control valve and the manual variable throttle valve.

The pressure-compensated flow-control valve

This type of valve is designed as an in-line valve with internal thread connections. A built-in pressure compensator ensures that the flow rate through the valve is kept constant and independent of the pressures in the inlet and the outlet connections. The desired flow rate is manually adjusted via the rotational position of a handle on the valve.

The valve is shown on Fig. 7.7. The valve block surfaces and the simple form enable simple and effective cleaning. The valve block is made of stainless steel ISO 316L and is acid-proof. The block material is in compliance with the FDA/3A regulations and is corrosion-proof against the hydraulic fluid and resistant to the cleaning means used within the food industry. The hydraulic fluid to be used is pure tap water (for water quality specification, refer to EU 80/778) or water added with 33% glycol (DOWCAL N).

Fig. 7.7 Picture of a Nessie® pressure-compensated flow-control valve, type VOH 30PM

In Fig. 7.8 a cross-sectional view of the pressure-compensated valve is shown. The valve contains two orifices. The flow-metering function takes place in orifice 1. The size of orifice 1 is manually adjusted by rotational positioning of the handle. For each position of the handle the corresponding pressure drop across orifice 1 is kept constant and independent of pressure variations in the inlet or outlet. This is accomplished by the compensation piston that regulates the size of orifice 2 according to the spring load and the inlet and outlet pressure. In order to ensure the function of the pressure compensation a preload of the spring is required, corresponding to a minimum pressure · drop of approximately 15 bar across the valve.

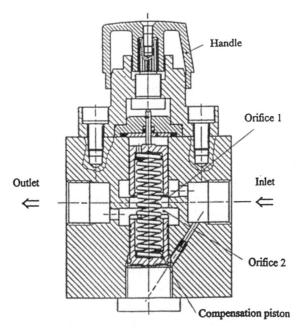

Fig. 7.8 A cross-sectional view of the valve shown in Fig. 7.7 (patent pending)

In Table 7.3 the most important technical data for the pressure-compensated flow-control valve in Fig. 7.8 are given.

Table 7.3

Technical data for a Nessie® flow-control valve,
type VOH 30M

Maximum flow	30 l/min
Maximum pressure drop	140 bar
Minimum pressure drop	15 bar
Minimum flow	2 l/min
Internal leakage	0.9–1.3 l/min
Temperature range	2–50 °C
Level of hygiene	High

A typical set of pressure-flow ($\Delta p - Q$) characteristics for a pressure-compensated flow-control valve is shown in Fig. 7.9. Note the required initial pressure drop across the valve and the robustness of the valve in keeping an adjusted volume flow rate constant over a wide range of pressure difference. The characteristics are close to the lines that are parallel with the pressure difference axis (compare with the ideal characteristic in Fig. 7.1).

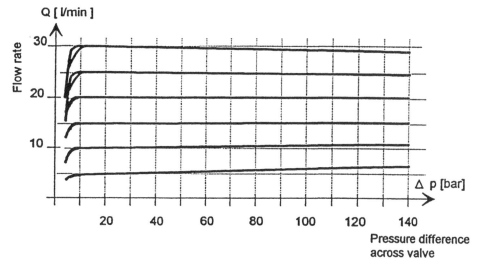

Fig. 7.9 Pressure-flow ($\Delta p - Q$) characteristics of a Nessie® pressure-compensated flow-control valve, type VOH 30M

Manual variable throttle valve

This type of valve executes very much the same function as the pressure-compensated flow-control valve described above. However, its functional principle is simpler. It is a seat valve and it contains no pressure compensation. The valve is also used for controlling the water flow rate and thereby the speed of an actuator. The desired flow is adjusted manually by setting of the rotational position of the handle on the valve. The flow can be adjusted across the full range of the valve. Further, the valve can be used as a completely tight shut-off valve. The valve is shown in Fig. 7.10.

Fig. 7.10 Nessie® manual variable throttle valve, type VOH 30M

In Fig. 7.11 a cross-sectional view of the variable throttle valve is shown. Special attention has been addressed to the detailed design of the valve stem and valve seat in order to solve the problems of erosion from cavitation and abrasion introduced by the high-energy density in water flow, the relatively high compression modulus (high stiffness) and the the relatively low damping (low viscosity) of water.

The design specifications, such as selection of non-corrosive materials and requirements of pressure fluid and surfaces, are equivalent to those used for the directional control valves and flow-control valves.

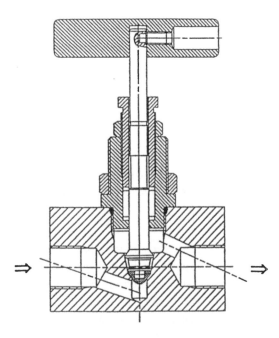

Fig. 7.11 A cross-sectional view of the valve shown in Fig. 7.10

In Table 7.4 technical data for a manual variable throttle valve are presented.

Table 7.4

Technical data for a Nessie® manual variable
throttle valve, type VOH 30M

Maximum flow	30 l/min
Maximum pressure drop	140 bar
Leakage (shut-off)	None
Temperature range	2–50 °C
Level of hygiene	High

In Fig. 7.12 a set of pressure-flow ($\Delta p - Q$) characteristics for a variable throttle valve as a function of the rotational position (N) of the handle is shown.

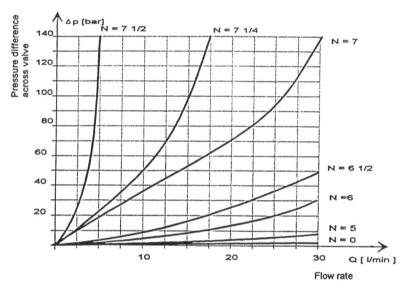

N is the number of rotations of the valve handle. At N = 0, the valve is fully open.

Fig. 7.12 Pressure-flow ($\Delta p - Q$) characteristics of a Nessie® manual
variable throttle valve, type VOH 30M

7.3 Pressure-control valves

Piston forces generated by linear actuators and motor torques generated by rotating actuators are functions of the working pressure in the hydraulic system. Precise control of the working pressure allows the system to perform any function where forces and torques have to be exerted between definite limits. Pressure-control valves maintain the working pressure in the hydraulic system in principle independently of variations in the fluid flow rate.

Two types of pressure-control valves will be discussed here: the pressure-relief valve and the safety valve.

The pressure-relief valve

The pressure-relief valve is used for protecting the components of a system against overload and for downloading the surplus water pressure flow from working pressure to a reservoir.

The valve is designed as an in-line valve with internal thread connections. The valve is shown in Fig. 7.13. the outside parts are made of corrosion-proof materials, stainless steel AISI 316L. The valve block surfaces and simple form enable simple and effective cleaning. The block material is in compliance with the FDA/3A regulations and is corrosion-proof against the hydraulic fluid and resistant to the cleaning means used within food industry. The hydraulic fluid to be used is pure water (water quality specification, refer to EU 80/778) or water added with 33% glycol (DOWCAL N).

Fig. 7.13 Nessie® pressure-relief valve, type VRH 60

In Fig. 7.14 a cross-sectional view of the pressure-relief valve is shown. The valve is designed as a simple, single-stage seat valve. The pressure set point (~opening pressure) is adjusted by preloading the spring acting upon the valve piston. A tube protects the spring against flow forces from the return water flow. Opposite the valve seat the piston end fits into a damping chamber. This built-in damping ensures a smooth motion of the piston. The internal valve details are all made of non-corrosive materials such as stainless steel, polymers and brass.

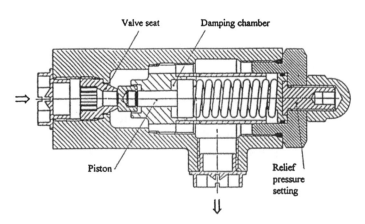

Fig. 7.14 A cross-sectional view of the valve shown in Fig. 7.13 (patent pending)

In Table 7.5 important engineering data for the pressure-relief valve shown in Fig. 7.14 are tabulated.

Table 7.5

Technical data for a Nessie pressure relief valve,
type VRH 60

Maximum flow	60 l/min
Pressure setting range:	
- Setting I	25–80 bar
- Setting II	80–140 bar
Dynamic response:	
- Overshoot @ flow step 30 l/min	8 bar
- Settling time @ flow step 30 l/min	25 msec
Internal leakage @ 112 bar	0.120 l/min
External leakage	None
Hysteresis @ p_{Max}	2.5%
Temperature range	2–50 °C
Level of hygiene	High

In Fig. 7.15 pressure-flow ($\Delta p - Q$) characteristics for a pressure-relief valve like the one shown in Fig. 7.14 are presented. Note that two different pressure setting ranges, I and II, are shown. Note further that the characteristics are lines approximately parallel with the flow rate axis (see the ideal characteristic in Fig. 7.1).

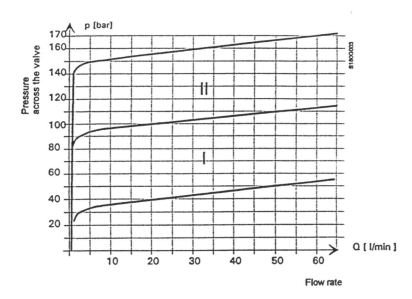

Fig. 7.15 Pressure-flow ($\Delta p - Q$) characteristics of a Nessie® pressure relief valve, type VRH 60

The safety valve

The purpose of a safety valve is to protect the hydraulic system against overload pressures and pressure peaks (transient pressures). The most important property of a safety valve is to execute dependable control of maximum pressure level. Safety valves are usually not adjustable but are adjusted to standard pressure settings. For the valve shown in Fig. 7.16a these settings are, respectively, 90 and 160 bar. Safety valves are not expected to operate unless a breakdown occurs.

General design specifications for the safety valve shown in Fig. 7.16a are similar to the pressure relief valve described in the previous section.

In Fig. 7.16b the $\Delta p - Q$ characteristics for the two standard pressure settings, marked I and II, are shown; they are similar to the characteristics in Fig. 7.15.

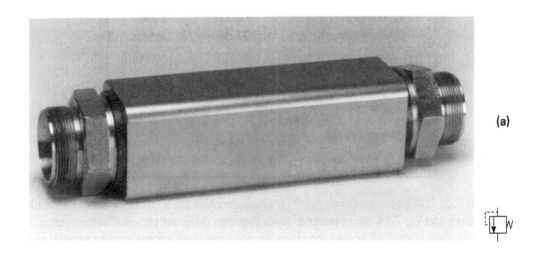

(a)

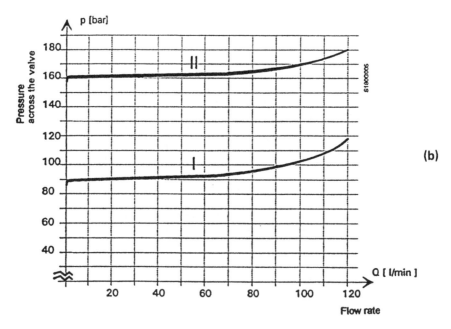

(b)

Fig. 7.16 (a) Nessie® safety valve, type VSH 20
(b) pressure-flow ($\Delta p - Q$) characteristics

7.4 Proportional control valves (ref. 21)

Water hydraulic proportional control valves and associated control electronics were introduced for practical applications some years ago. Typical applications are pressure controls for large presses, mostly within the steel industry.

Since water hydraulics may gradually spread into applications similar to those for oil hydraulics, the demand for increased performance of water hydraulic proportional valves and associated electronics is increased.

Proportional control in water hydraulics of pressure and flow rates has been implemented as a flow inlet/outlet device introduced by the firm Hauhinco and schematically illustrated in Fig. 7.17. Two proportional valves of a 2-port, 2-way type (position No. 3), are used for the control of a single-acting cylinder (position No. 1). One valve controls the flow rate into the system, and the other valve controls the flow rate out of the system. The valves are continuously activated via solenoids energized from a control unit (position No. 4). This unit contains a digital electronic control circuitry to achieve efficient position, velocity and pressure control, e.g., via a PID (*p*roportional, *i*ntegral, *d*erivative) control law and on the basis of relevant feedback signals from the transducers (position No. 2).

Position No. 6 indicates a directly operated ceramic 3-port, 2-way proportional spool valve currently under development. This valve will replace the two 2-port, 2-way valves marked by position No. 3. The valve at position No. 5 is used only with the inlet/outlet system to ensure a defined power output or fail-safe position for the main actuator/user device (position No. 1).

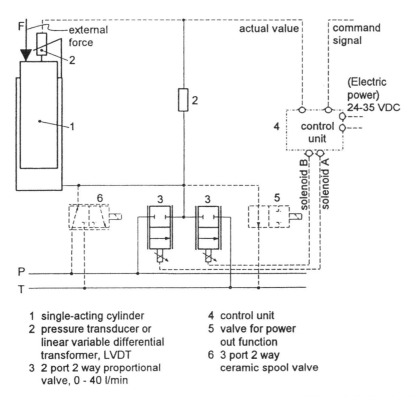

1 single-acting cylinder
2 pressure transducer or
   linear variable differential
   transformer, LVDT
3 2 port 2 way proportional
   valve, 0 - 40 l/min

4 control unit
5 valve for power
   out function
6 3 port 2 way
   ceramic spool valve

Fig. 7.17 Water hydraulic proportional control valves used in an inlet/outlet device
(courtesy Hauhinco)

8. POWER SUPPLIES

The hydraulic power supply, often called the hydraulic power unit, serves a central function in the whole hydraulic system. The power unit generates the hydraulic fluid flow and fluid power to be used in the system. Basically all industrial hydraulic power units are pretty much the same, although there are vast differences among the components applied in many types of units.

The basic unit may contain a hydraulic pump, a hydraulic reservoir (tank) with a cover, a suction strainer, a return filter, a motor coupling, an electric motor, a relief valve, a pressure gauge, hydraulic fluid and the necessary internal piping (see Fig. 8.1).

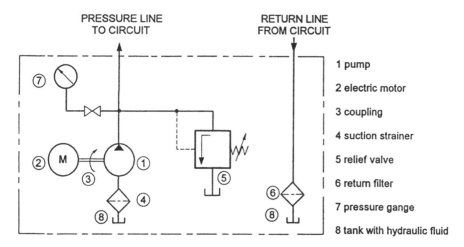

1 pump

2 electric motor

3 coupling

4 suction strainer

5 relief valve

6 return filter

7 pressure gange

8 tank with hydraulic fluid

Fig. 8.1 Schematic diagram showing the elements of a hydraulic power unit using a pump with constant displacement

8.1 Purpose and dimensions of the reservoir

A well-designed reservoir serves some important functions in the hydraulic power unit. The most important of these include the following:

(1) The reservoir must be large enough to store more than the largest volume of hydraulic fluid that the hydraulic system will require. Note that when actuators move, the fluid level in the reservoir will change. Also leakage in the system should be accounted for.

(2) The reservoir must also provide a volume of fluid large enough so that air and vapor bubbles and particles of contaminant material can be separated out. Because the mass density of water is approximately 15% higher than the mass density of mineral oil, the settling path for particles will be longer in a water tank than in a mineral oil tank. As a rule of thumb, the hydraulic fluid volume in a stationary water tank should be three to five times the hydraulic volume flow rate in volume flow per minute.

(3) In a hydraulic system power losses in the components due to friction and leakage will occur. These power losses will accumulate as heat in the hydraulic fluid. The reservoir should help dissipate such heat. Because of the relatively high vapor pressure of

water, the water temperature should be kept below 50°C. This necessitates a water tank of larger volume compared with an equivalent mineral oil tank.

The many requirements of good reservoir design apply to most types of reservoirs. The following are some important guidelines.

The location of lines. The intake of suction lines should be well below the level of the hydraulic fluid but not closer to the tank bottom than 1.5 times the pipe diameter in order to avoid intake of contaminants settled at the bottom.

Return lines should terminate below the level of the hydraulic fluid in order to prevent aeration of the fluid. The return line outlet should direct the flow towards the wall tank at an angle of >45°. The return outlet should be clearly separated from the suction intake at a distance as great as possible by a vertical baffle plate. This ensures a long flow path for the hydraulic fluid, whereby "whirlpooling" is avoided, dirt is separated out, and cooling is improved.

Gravity drains from hydraulic components with seal or spring cavities should enter the tank above the fluid level and be separated from return lines.

Cleaning provisions. In a hydraulic system there are two sources of contaminants: those that are introduced by the operation of the system and those that are left from the manufacturing of the reservoir. Standard cleaning procedures defined by the system supplier should be followed closely. In case of doubt, the supplier must be consulted before start-up of the system.

The tank containment should be designed in a way that makes the inside of the tank easily accessible through a hatch.

Draining. For periodic cleaning of the tank, provisions should be made for emptying the tank, such as via a tank drain tap at the bottom.

Filling. For filling the tank a filling valve should be installed.

Level indicator. The hydraulic fluid level is critical for proper functioning of the hydraulic system. A gauge indicating high level, nominal working level and low level should be installed.

Air filter. The hydraulic fluid level in the tank varies during the hydraulic system's operation. Therefore the air space above the fluid in the tank must breathe, i.e., sometimes air from outside has to be sucked in or air from the inside has to be pressed out. This air flow takes place through an air filter, a so-called air breather. The capacity of such a filter must be adequate for an air flow rate equal to the hydraulic pump flow rate.

Heat exchangers. Often even a well-designed reservoir cannot sufficiently dissipate the heat from the power losses in the hydraulic system, in which case the temperature of the hydraulic fluid under a steady-state operation will not remain below an acceptable level **(for pure water this should normally be no higher than 40°C)**. In such cases the hydraulic fluid is cooled in a separate cooler, e.g., a water cooler or an air-blast cooler. In Fig. 8.2 a typical temperature curve during a duty cycle for hydraulic fluid in the reservoir is shown. Sometimes the hydraulic fluid needs to be heated up to an acceptable temperature before the hydraulic system can start up in normal operation. Various types of heating units may be available. For water hydraulic systems using pure water the starting temperature for the water pressure medium must be above 3°C.

The use of heat exchangers, coolers or heating units, may permit a reduction in the reservoir size. In many cases the heat exchanger may be a separate module in the hydraulic power supply, sometimes equipped with its own separate circulation pump with thermostat control. Stainless steel should used in heat exchanges, coolers and heating units.

Filter access. If the pump suction filter is submerged in the reservoir, access should be provided so that it is possible to service the filter without draining the reservoir.

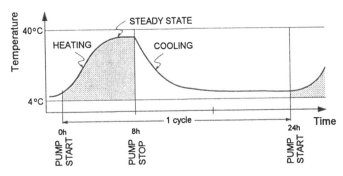

Fig. 8.2 Temperature curve for the hydraulic fluid in a reservoir
with an operational period of 8 hours once every 24 hours

8.2 Tank layout

In general the hydraulic reservoir in a hydraulic power transmission system may be more or less integrated with the machine or plant that it serves. However, in stationary industrial-type applications where space is not limited, a separate hydraulic reservoir may be the optimal choice. A separate reservoir can be configured in various ways, according to specific needs.

A classical configuration is shown in Fig. 8.3. The reservoir is rectangular, with the pump and electric motor mounted on the top. This provides an efficient and compact construction: the pump suction line is reasonably short and the tank can be opened for any internal service needed, such as cleaning of the strainer on the pump suction line. The tank top must be of structurally rigid design in order to support the pump and the electric drive motor. Often these are mounted on a separate plate attached to the tank top by resilient mounts. This arrangement helps dampen vibrations and noise, and assists in heat dissipation.

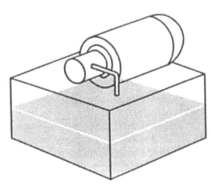

Fig. 8.3 Rectangular reservoir (tank) with pump and electric motor
mounted horizontally on the tank top

The pressure in the pump intake line is critical for avoiding cavitation in the pump. As an example, the pressure relationship in a suction line of a hydraulic water pump is illustrated in Fig. 8.4. The various pressure losses contributed by the individual elements of the suction line are added together and shown in the graph.

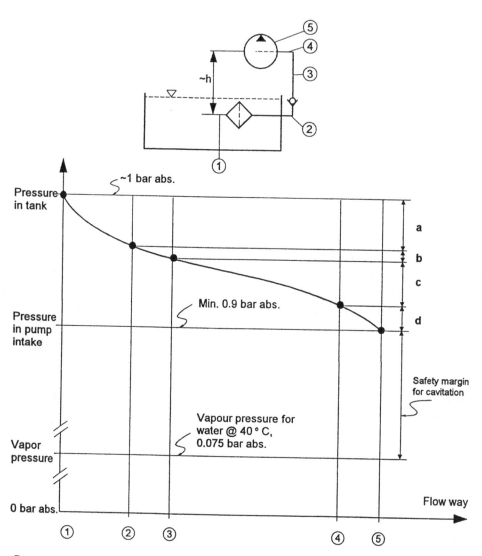

Pressure loss:

a Filter (1-2)
b Valve (2-3)
c Pipe (3-4)
    (loss due to resistance,
    velocity*, height h)
d Pump entry (4-5)

* It is recommended that the flow velocity in the pump suction line be less than 1,5 m/sec

Fig. 8.4 Pressure losses in a pump suction line

In the arrangement shown in Fig. 8.5 the pump and motor are mounted vertically. The reduced space offers an economical solution. The arrangement provides maximum protection against physical damage to the pump and reduces suction height requirement. Service of the pump becomes a little more difficult than in the arrangement shown in Fig., 8.3 especially when large pump units are used.

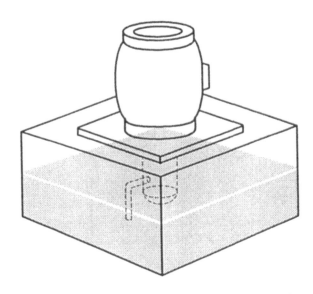

Fig. 8.5 Rectangular reservoir with pump and electric motor
mounted vertically on the tank top

In Fig. 8.6 an arrangement is shown where the pump is mounted adjacent to a relatively narrow, tall rectangular tank on a common base frame. A design in which the suction line enters below the fluid level in the tank provides a positive head to the pump but still allows access for the maintenance of the pump. In order to permit service on the pump without draining the tank, a shut-off valve must be included in the suction line.

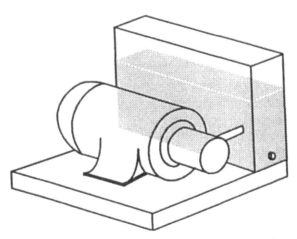

Fig. 8.6 Rectangular, narrow reservoir with pump and electric motor
adjacent to the tank in an L-shaped arrangement

In Fig. 8.7 the tank is placed above the pump and the motor. This provides excellent inlet conditions to the pump. A shut-off valve, however, must be included in the suction line so draining of the tank is not necessary when the pump is serviced. Accessibility to the pump is good, although service space is somewhat limited.

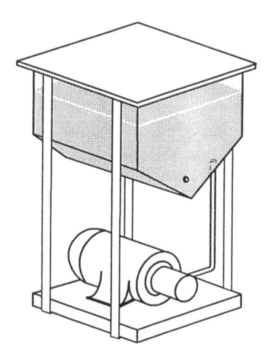

Fig. 8.7 Reservoir suspended above pump and electric motor

For economical solutions, standard commercial hydraulic power packs are available on the market but with a limited set of features. In order to meet customers' special requirements, some suppliers offer supplementary modules as options to standard units.

In Fig. 8.8 a water hydraulic Nessie® power pack with a water/water cooler in the standard version is shown. The tank size is 250 l and the pump delivery is up to 112 l/min. The pump and the electric motor are connected with a bell housing. The motor is fastened to the water tank by screws in a base plate. The pump is driven via a flexible coupling. The water tank is made of welded stainless steel plate and is accessible via a hatch on the side. A tank made of polymer plate is also available. A water level sight glass allows for the inspection of the water level. A pressure relief valve is built into the pump pressure line. The valve opens to the tank if the pre-set pressure is exceeded. The pump suction line is connected to the tank via a hose. A return filter is built into the return line to ensure that return water to the tank is filtered as appropriate for the chosen filter size. The return line is also fitted with an air breather which ensures that air entering the tank is filtered or that air pressure in the tank is relieved.

As mentioned above, many hydraulic power units may be custom designed in order to meet specific requirements. An example of this is the L-shaped Nessie® water hydraulic power pack in a base plate version illustrated in Fig. 8.9.

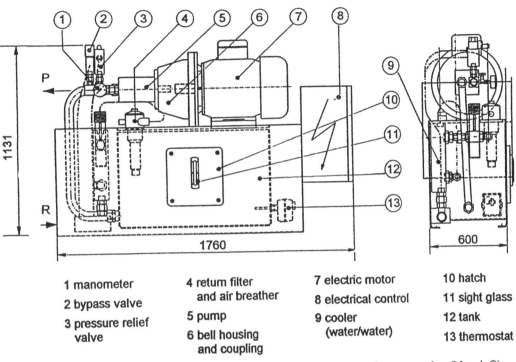

1 manometer	4 return filter and air breather	7 electric motor	10 hatch
2 bypass valve		8 electrical control	11 sight glass
3 pressure relief valve	5 pump	9 cooler (water/water)	12 tank
	6 bell housing and coupling		13 thermostat

Fig. 8.8 Water hydraulic power pack with 250-l tank and water/water cooler (Nessie®)

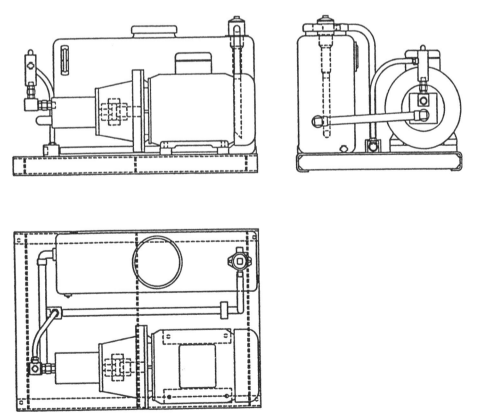

Fig. 8.9 Custom-built power pack with pump mounted to one side of the tank (Nessie®)

8.3 Filters

Filtration in general

In any hydraulic system it is of extreme importance that the hydraulic fluid fulfill certain specified minimum requirements of cleanness to ensure the correct functioning and reliability of the system and the expected lifetime of the components.

Any hydraulic system is to some extent contaminated by dirt. To keep impurities from contaminating the hydraulic fluid, the hydraulic system must therefore contain a specified filter (or filters) for continuous filtering of the fluid during the operation of the system. A correct filtering process should remove from the hydraulic fluid those solid contamination particles, often called dirt, which by size, shape and composition may interfere with or harm the system or its components.

Dirt may build up across small control orifices and thereby impede or prevent their control function. Dirt in narrow clearances between moving parts may cause these to stick. These phenomena may happen to valves with sliding spools, where dirt in the hydraulic fluid can cause the spool to be jammed. Dirt may also hinder lubrication and lead to excessive wear between moving parts. Dirt may cause erosion, e.g., when particles in hydraulic high velocity jet flows slide along the walls in control orifices or when the jet flow strikes a control wall in the valve. Finally, dirt may lead to bacterial and other microbial growth, especially when the hydraulic fluid is water-based.

There are many sources from which foreign material, particles, etc., can contaminate the hydraulic fluid: chips from machining the component parts, dust from the atmosphere, wear particles from the components, dirt from the fluid itself, etc.

Theoretically speaking, a filter should remove all solid contaminant particles from the hydraulic fluid. In practice, however, complete filtration of the minutest particles would lead to insurmountable costs.

Contamination classes and filtration ratings

In relation to filter sizing particles are measured in 10^{-6} m = μmeter (μm) = microns. For comparison, the diameter of a human hair is 70–80 μm, the resolution of the human sight is 40–50 μm, and the diameter of bacteria is 2–3 μm.

With respect to the cleanliness of hydraulic systems, several classification or contamination codes for particles have been established, for example, NASA (National Aeronautics and Space Administration, U.S.) standards and the ISO/DIS 4406 standard. A new, clean hydraulic oil often may comply with class 9 requirements of the NASA standard contamination code (see Table 8.1).

As expressed above, it is uneconomical and impractical to remove all particles (dirt) from the hydraulic fluid. Therefore, optimal filter sizing must be established. The closer the maximum size of the particles in the hydraulic fluid comes to the size of critical gaps in the components, the more probable a jamming situation will occur. An excess of smaller particles tends to cause wear failures by increasing leakage and decreasing efficiency and performance.

A rule of thumb claims that the absolute filtration should equal one half of the smallest clearance between moving parts of the system components. Filter ratings are normally given in microns.

Table 8.1

Class 9 in NASA standard contamination code. Size range (in μm)
for number of particles per 100 ml sample of hydraulic fluid

Range (μm)	Maximum number of particles
5–15	128,000
15–25	22,800
25–50	4050
5–100	720
> 100	128

In Table 8.2 typical clearances of various hydraulic components are shown. The values shown are an average guide. The actual critical clearance values in a given application should be obtained from the component manufacturer.

Filter ratings are normally defined in microns. The degree of filtration is expressed in terms of nominal and absolute filtration rating.

Table 8.2

Typical clearances in hydraulic components

Component	Clearance (μm)
Vane pump	
Tip of vane	0.5–1
Sides of vane	5–15
Piston pump	
Piston to bore	5–40
Cylinder to valve plate	0.5–2
Gear pump	
Side plate	1–100
Gear tip to housing	2–100
Actuators	5
Control valve (valve spool/sleeve)	1–23
Servovalve	
Orifice	130–450
Valve spool /sleeve	1–4
Hydrostatic bearing	1–25

The National Fluid Power Association (NFPA) in the United States has proposed the following definitions:

Absolute filtration rating: The diameter of the largest hard spherical particle that will pass through a filter under specified test conditions. This is an indication of the largest opening in the filter element.

Mean filtration rating: A measurement of the average size of the pores of the filter medium.

Normal filtration rating: An arbitrary value (in microns) indicated by the filter manufacturer. Due to lack of reproducibility this rating is deprecated.

The usual practice is to rate filters both in nominal and in absolute values. Frequently the filtering characteristics are expressed with the β ratio. This ratio indicates the efficiency of contaminant removal from the hydraulic fluid flowing through the filter elements and is defined by the ratio:

$$\beta_x = \frac{\text{Number of upstream particles larger than } x \text{ } \mu m}{\text{Number of downstream particles larger than } x \text{ } \mu m}$$

This means that a filter with the rating $\beta_5 = 10$ will remove 90% of the particles equal to or greater than 5 μ. The filter rating of $\beta_x = 75$ is normally considered to be equivalent to the absolute filtration rating, i.e., that 98.7% of the particles equal to or greater than x μ are removed from the hydraulic fluid passing through the filter.

Filter location

In general there is no best location of a filter in a hydraulic system. Sometimes, especially in high-performance systems, two or more locations are preferred.

For the following description of various filter locations, refer to Fig. 8.10.

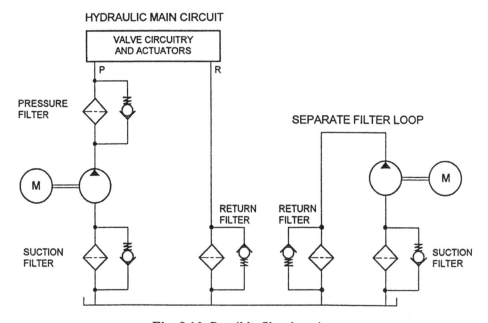

Fig. 8.10 Possible filter locations

Suction line filter: This filter is placed at the intake of the suction line in the tank between the tank and the pump, and it primarily protects the pump from large particles. In order to prevent cavitation in the suction line or pump, the pressure drop across the suction filter should be kept low (<< 0.1 bar) (see Fig. 8.4).

Often the terms **"suction filter"** and **"strainer"** are used interchangeably. Even though their functions are very much alike, a strainer is a device that removes solid particles moving in a straight line into the strainer, and a filter is a device that removes solid particles moving along a tortuous path through the filter, where they may be constrained by the filter element.

Pressure line filters: This filter is placed in the pressure line right after the pump. If it is placed before the relief valve, it filters the total pump delivery.

Sometimes the pressure filter is placed in a bypass arrangement. The filter is then placed in a line parallel to the pressure line and the flow through the two lines divides inversely to the pressure drop in each line. In the bypass arrangement only a fraction of the full flow rate will be filtered.

Return line filter: This filter is located in the main return line. It provides full-flow filtration, but the filter is not subjected to withstand high pressures.

Separate loop filter: This filter arrangement requires a separate pump and a power drive. It provides a continuous filtration, independent of the flow in the main circuit. The separate loop may be used for additional purposes, such as for circulating fluid to a cooler system or for boosting of the main pump.

Types of filters

A large variety of filter makes is available on the commercial market. The selection of the right filter for a given application is a matter of matching filter characteristics with the requirements and of using best practice. Note that choosing a wrong filtering technique may ruin the hydraulic system.

All filters are built of media acting to varying degrees as both depth and surface filters.

Surface filtering. By surface filtering (see Fig. 8.11) the contamination particles are removed by the flow of the hydraulic fluid through a two dimensional surface containing a uniform distribution of uniform orifices. The particles larger than the orifices will be caught by the surface filter, while the smaller particles will pass through.

Surface filter material may be wire mesh, i.e., wires with small uniform diameter that are woven into a square or other regular geometrical pattern.

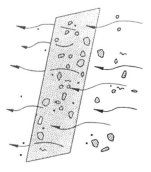

Fig. 8.11 Surface filtering

Depth filtering. By depth filtering (see Fig. 8.12) the contamination particles are removed by following a tortuous path through the filter media. Depending on the filter rating, the larger particles will be caught in the filter along the three-dimensional path through the filter media.

Depth filter material may be **fibers**, for example, paper, cellulose, felt or fiberglass; or **sintered powder**, for example, metal, ceramics or plastics.

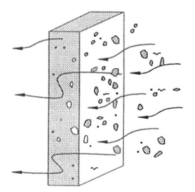

Fig. 8.12 Depth filtering

Filter capacity. A rule of thumb recommends that the filter capacity in flow volume per unit time should be equal to two times the pump delivery (in flow volume per unit time). Another, more accurate rule may be to select a filter for which the maximum pressure drop across the filter is acceptable for the hydraulic system in question.

Practical filters

In Fig. 8.13 a Nessie® suction strainer is illustrated. The strainer is used in water hydraulic systems for cleaning the water before it is sucked into the pump. The strainer is placed below water level, at the intake of the suction line in the tank. The star-shape cross-section ensures a large capacity (max. 150 l/min) with a minimal pressure drop (<0.03 bar @ max. flow rate). The filtration rating is 100 μ. The material is stainless steel or aluminum.

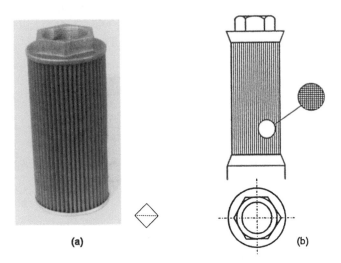

(a)   (b)

Fig. 8.13 Suction strainer: (a) photo, (b) design detail

In Fig. 8.14 a Nessie return filter is shown. The return filter is designed especially to ensure trouble-free operation in Nessie water hydraulic systems. The filter is made of corrosion-resistant materials such as reinforced plastics and stainless steel. The filter element consists of inorganic glass fibers. It is installed in the main return line and combined with an air breather. In addition the filter serves as a filling filter. To show dirt accumulation the filter can be fitted with a manometer and/or an electrical on/off pressure switch.

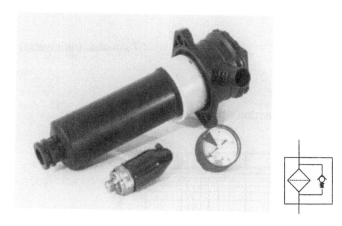

Fig. 8.14 Return filter (Nessie®)

In Fig. 8.15 a schematic cross-section of a return filter of the type in Fig. 8.14 is shown. The filter media is a depth-type filter element.

The technical data of a Nessie return filter of type FRH is shown in Fig. 8.16. The filtration rating, β ratio, and the pressure drop as a function of flow rate are shown graphically.

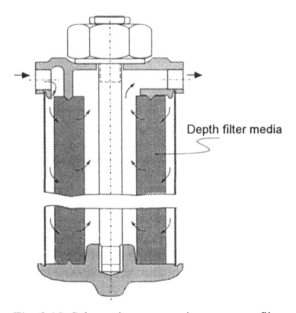

Depth filter media

Fig. 8.15 Schematic cross-section or return filter

Return filter
- Flow: Max. 70 l/min.
- Filter fineness: 10 μm (β_{10} = 75) (see diagram below)
- Pressure drop: See diagrams below
- Dirt capacity α (ACFTD):   7.5 g
- Bypass valve: Opens at 2.5 bar
- Connection: ø20.5 mm
- Filter material: Exapor, inorganic glass fiber

Breather
- Filter fineness: 7 μm abs. (renewable)

Filtration rate
Filtration quotient β in function of particle size X (ISO4572-81).

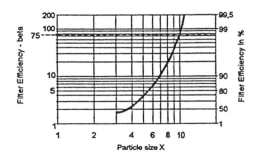

Pressure drop
Pressure drop, in bar, over the filter in function of flow (l/min.)

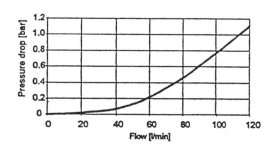

Fig. 8.16 Technical data of a Nessie® return filter of type FRH

8.4 Control and monitoring of system

In Fig. 8.17 a standard monitoring and hydraulic control system for a Nessie water hydraulic power pack (of the type in Fig. 8.8) is shown along with the various components in the system.

In Fig. 8.18 a supplementary control module and its various components are shown. Other extensions and options are, of course, possible.

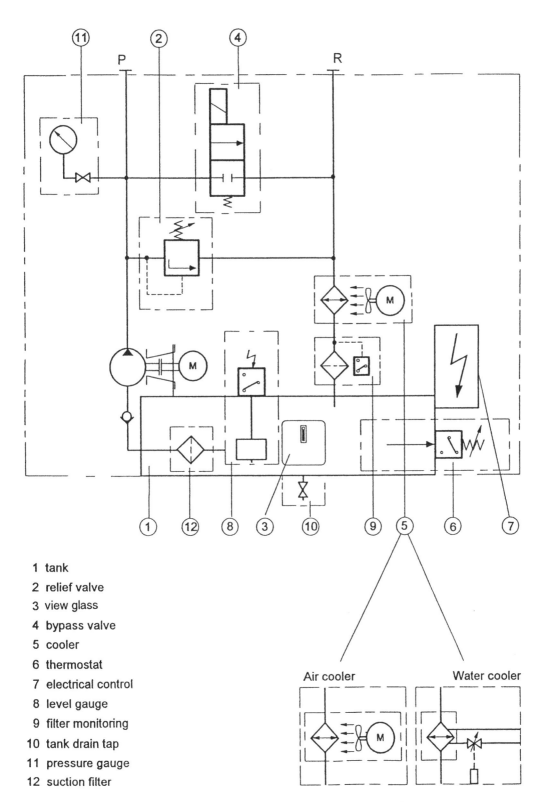

Fig. 8.17 Reservoir monitoring and hydraulic control system

1 tank
2 relief valve
3 view glass
4 bypass valve
5 cooler
6 thermostat
7 electrical control
8 level gauge
9 filter monitoring
10 tank drain tap
11 pressure gauge
12 suction filter

Air cooler   Water cooler

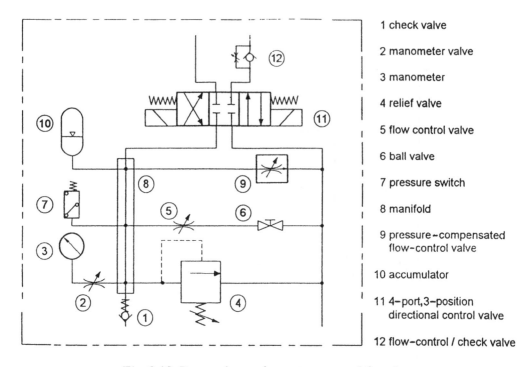

Fig. 8.18 Reservoir supplementary control functions

8.5 Accumulators

Accumulators are devices used to store energy. The energy is stored in the form of hydraulic fluid volume under pressure. The applications of accumulators are numerous. Typical examples include the following:

Supplementary pump delivery. On machinery containing a multiple set of operations, where the working cycle requires a peak requirement of the hydraulic fluid flow rate, an accumulator may substitute for a large and costly pump. It is necessary only that the pump have a delivery averaged over a continuous operating cycle period that is larger than the averaged flow rate demand. The accumulator is then filled during the operating cycle when the flow demand is low, or during idling intervals.

Maintaining pressure. In hydraulic systems an accumulator can compensate for leakage over a certain period of time and thereby maintain pressure, for example, in an activator exerting a clamping force. While the accumulator is active, the pump delivery is circulated to tank via an unloading valve, or the pump may supply another system or be stopped altogether.

Absorbing shock. Sudden valve closure may cause pressure peaks and shock waves (water hammer). Likewise, pressure shocks may result from external mechanical impacts on cylinders or motors. Appropriate sizing and placement of accumulators are used for partial damping of such pressure transients.

Damping of pump delivery pulsations. The displacement of most hydraulic pumps (positive pumps) varies as a function of the pump shaft rotation. This will introduce pulsations in the pump delivery. Depending on the properties of the hydraulic system supplied by the pump and on the amplitude of pump pulsations, a small accumulator positioned upstream and "looking" directly into the pump outlet may act as a pulsation damper.

Accumulators may be divided in three categories: weight-loaded, spring-operated and air pressure–operated.

The weight-loaded accumulator

The weight-loaded accumulator consists of long, vertically mounted piston loaded with a heavy weight (see the schematic unit in Fig. 8.19). The piston can move in the cylinder and exert a pressure on the hydraulic fluid. The weight can be any type of heavy material. It is important that the piston and cylinder have an accurate fit so that leakage is minimal. This type of accumulator provides a constant static pressure, but it is heavy and costly. It is also slow to respond when quick changes in flow demand in the system are required. Due to its low damping properties, water-hammering effects are undampened.

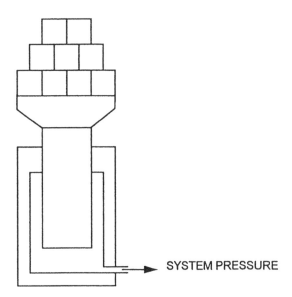

Fig. 8.19 Schematic of weight-loaded accumulator

The spring-operated accumulator

Usually spring-loaded accumulators are much smaller and cheaper than weight-loaded accumulators. The spring-loaded accumulator (see Fig. 8.20), consists of a pre-loaded spring acting upon a piston in a cylinder. The piston exerts a pressure on the hydraulic fluid. The energy stored in the spring is converted to hydraulic pressure and flow. Practical applications are mostly medium pressures and small fluid flows. The pressure varies with the spring compression.

The bladder-type accumulator

The bladder-type accumulator consists of a bladder or a bag of synthetic, elastic material that is pre-charged with nitrogen to a predetermined pressure (see Fig. 8.21). The bladder is mounted in the accumulator containment shell, and the space outside the bladder is filled with hydraulic fluid. When fluid is pumped into the shell from the pump, the gas in the bladder is compressed and a reservoir of pressure energy is built up. When pressurized hydraulic fluid is called for, the bladder will expand and deliver hydraulic energy, fluid and pressure, to the hydraulic circuit. The pressure in the bladder accumulator will vary according to the initial nitrogen pre-charge of pressure and the volume of the bladder.

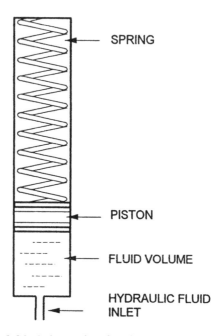

Fig. 8.20 Schematic of spring-type accumulator

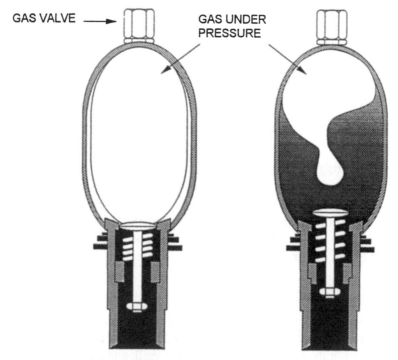

Fig. 8.21 Bladder-type accumulator

Examples of accumulator applications

In Fig. 8.22 a hydraulic circuit uses an accumulator for compensation of leakage in the system. The piston exerts a given force on a clamp device, and the pressure in the accumulator unloads the pump and maintains the clamping pressure. When the pressure due to leakage becomes too low, the unloading valve closes and the pump will re-establishes the pressure.

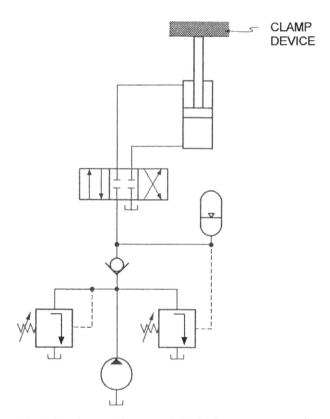

CLAMP
DEVICE

Fig. 8.22 Accumulator used for leakage compensation

In Fig. 8.23 pulsation damping of a hydraulic pump is shown. A small accumulator is mounted so it "looks" directly into the pump pressure line. The accumulator aims at smoothing out the pump pulsations by delivering hydraulic fluid flow close to 180° out of phase with pump flow peaks.

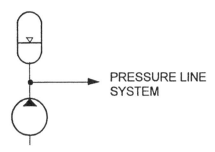

PRESSURE LINE
SYSTEM

Fig. 8.23 Pump pulsation damping with small accumulator

9. DESIGN OF WATER HYDRAULIC SYSTEMS

As described in chapter 1 hydraulics has long played an important role in the design and development of machines and production plants. Without the application of hydraulic components and systems the present state of mechanization and automation could properly never have been achieved. The advantages of hydraulics are numerous. In a hydraulic system the ratio of power to weight and power to volume is relatively high. Linear motion is very easy to establish with hydraulic cylinders and the speed of hydraulically powered pistons can be precisely controlled over a wide range of speeds (more than 1:100). In hydraulic systems high forces and torques can easily be controlled over the full range of speeds and can be transmitted over small or medium distances Overloads can be safely controlled.

Originally, the only available hydraulic fluid pressure medium was water. Due to many design problems in pumps, valves, etc., mineral oil became the more popular pressure medium because the higher viscosity of oils offers advantages such as better lubrication between moving parts and a good leakage control in the hydraulic components. Furthermore, mineral oil helps minimize corrosion of traditional materials.

In recent years by taking advantages of using new types of materials and new component design principles, the traditional problems of water hydraulics have been overcome and water has again become attractive as a pressure medium because it offers unique advantages such as environmental friendliness and no risk of fire. These advantages open up many application areas for water hydraulic power transmission.

Before the hydraulic design engineer is presented with a specific hydraulic design task, a higher level design decision has normally already been made as to whether an electro-mechanical, pneumatic, water hydraulic or oil hydraulic solution should be considered for the power drive system for a machine design project. The task assignment of the hydraulic design engineer is then to work out an optimal hydraulic solution for the drive.

An optimal solution is by no means unambiguous, because a sensible compromise among several design criteria must be made. Such criteria are:

- Minimum space requirements
- Easy integration of the system with the machine design project
- High accuracy
- Cost-effectiveness
- Reliability
- Minimum energy consumption
- Minimum maintenance
- Easy maintenance
- Use of standard components
- Low environmental impact
- Low fire hazard

The optimal hydraulic solution must typically meet the expectations and requirements of three principal parties: (1) the customer, i.e., the machine developer, (2) the production manager or party responsible for acquiring a reliable, hydraulically controlled machine on schedule and within budget, and (3) the operator, who needs a system with low and predictable maintenance.

The starting point for achieving an optimal hydraulic solution for a machine power drive is to obtain a set of precise functional specifications of forces (torques) and speeds for the planned actuators and learn the requirements for the operational sequence times of the machine to be developed. On this basis a hydraulic circuit is designed and the components are selected or designed. To accomplish this, the following considerations should be observed:

(1) Whether hydraulic control of the machine drive in question is to be designed with water or oil hydraulics, the functional circuit layout and the dimensioning procedure for the system are, in principle, identical. The various differences in physical properties between water and mineral oil, however, should be taken into account. The selection of which type of drive system to chose–water hydraulics or another category–for a specific application is often an a priori decision made at a higher project management level.

(2) Components laid out and designed for oil hydraulic applications **can generally not** be applied in a water hydraulic system because of problems with corrosion, lack of lubrication and internal leakages. In case of doubt the component supplier should be consulted!

(3) Water hydraulic systems and oil hydraulic systems have several quite different maintenance and service requirements. These have to be carefully noted and taken into account before a water hydraulic system is acquired and put in operation.

9.1 System specification

The specific requirement of a hydraulic system is to ensure that a load is moved or carried in a specified way, or expressed more precisely: the hydraulic system must interact with the load such that the position, speed, acceleration and/or force progress in time as specified.

With reference to Fig. 9.1 the general way in which the hydraulic system achieves purpose can be explained as follows: mechanical power in the form of torque-speed inputs to the hydraulic pump is transformed into hydraulic power in the form of pressure-flow rate states in the hydraulic system. These pressure-flow rate states are further transformed into torque-speed outputs of the rotary hydraulic motor or, in the case of a linear hydraulic motor (a cylinder), into force-speed outputs.

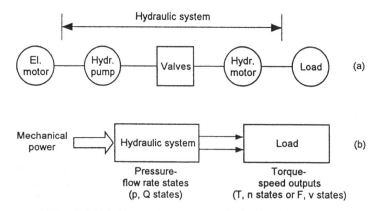

Fig. 9.1 (a) General structure of a hydraulic system
(b) The power transmission from the hydraulic system to the load

The hydraulic system must, of course, be dimensioned such that it can overcome the power requirements from the load and generate the required speeds and torques (forces).

As the basis for the engineering dimensioning of a hydraulic system the following two main categories of specifications must be defined and agreed upon by the machine design project team (or with the customer): (a) torque, speed and power specifications and (b) position and sequence specifications.

Torque, speed and power specifications (ref. 19)

These specifications must be worked out for each required actuator (motor). A stringent way to do this is by formulating a load diagram, often termed a **load locus,** for the actuators. Such a diagram defines the relationship between external load forces and required actuator speeds for linear motors (cylinders), and also between external load torques and required rotational speeds for rotary motors. Typical load-locus diagrams for mechanical friction load force and mechanical harmonic vibration forces for mass or spring loads are illustrated in Figs. 9.2 and 9.3. Note that the loci in Fig. 9.3 become circles by adjusting the scales of the coordinate axes.

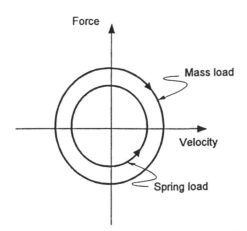

Fig. 9.2 General mechanical friction load (ref. 2)

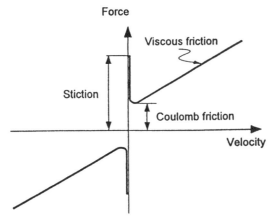

Fig. 9.3 Mechanical harmonic vibration loads (refs. 2 and 18)

Often the anticipated loads for a hydraulic system are only vaguely specified by the user, and therefore the hydraulic system designer may have to estimate the load diagrams. In Fig. 9.4 a complete load locus for a single actuator is shown. Required force-speed states are plotted (x) in the diagram. It is especially important to identify required maximum force (F_{MAX}) and maximum speed (v_{MAX}). The curve (envelope) from the F axis to the v axis in Fig. 9.4, which demarcates the area containing points of required force-speed states, is defined as the **load locus**.

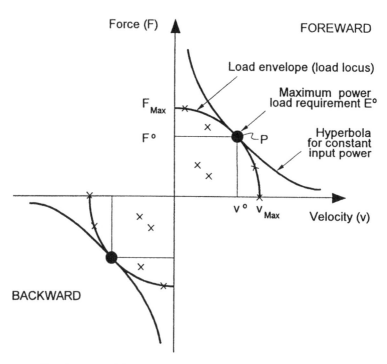

Fig. 9.4 Load locus for a linear motor (refs. 2 and 18)

The point P in Fig. 9.4, where the load locus is tangent with the hyperbola describing constant input power to the load, defines the maximum power $E^0 (= F^0 \cdot v^0)$ required by the load.

Fig. 9.4 illustrates a situation where the load resists the motion. This is called a positive load. The direction of the motion is, by definition, in a forward direction, with respect to the actuator, and therefore the load locus is shown in the upper right quadrant of the diagram. Likewise, having a positive load in the backward direction gives a load locus in the lower left quadrant. If loads become negative, i.e., when the loads pull the actuator, the load loci are located in the lower right or upper left quadrants.

Position and sequence specifications

Very often a hydraulic system consists of a number of different actuators, such as rotary motors and cylinders, that have to function in a required pattern of positions and motions in a timely, coordinated sequence in order to control the operation of an automatic machine or production system. This sequence must be uniquely defined and specified a priori to the hydraulic system designer before he or she can initiate the design work.

   The sequence can be clearly specified in a functional diagram (see Fig. 9.5), which schematically illustrates the relative motions and positions as a function of time for the different actuators. The function-time diagram for a double-acting cylinder is presented by a graph of its position-time relationship. For a motor the diagram presents the motor speed as a function of time. In Fig. 9.5 a function-time diagram for three cylinders and one rotary motor is shown. The diagram also specifies the required position and the timing signals for the actuation of the motors. By using comparably scaled coordinate axes, the cycle time of a repetitive sequence can be derived and the slopes of the actuator position-time functions can be determined to help define the linear actuator speeds. The diagram can be extended to specify other required time events in the sequence.

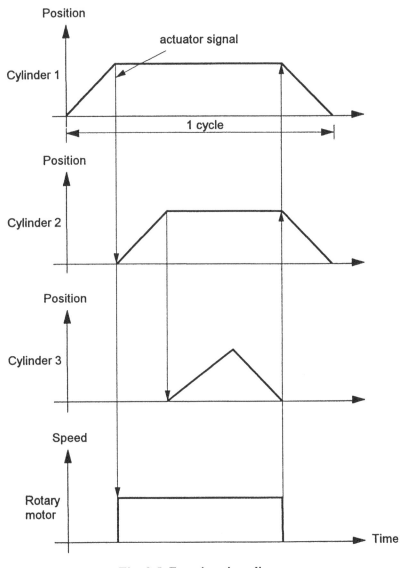

Fig. 9.5 Function-time diagram

Further specifications

Before one can begin to design of a hydraulic system, can begin a complete description of all the functional requirements must be available: operational requirements such as intermittence factors, integration with other types of machinery, availability of maintenance, and such conditions as system location and environmental requirements (humidity, noise, dirt, heat, etc.).

In addition to the above specifications there are, for some applications, requirements for dynamic performance such as bandwidth, positional accuracy and resolution. However, a further discussion of these is beyond the scope of this presentation.

9.2 Circuit layout

Any hydraulic system above a certain level of complexity should be documented in a hydraulic circuit diagram showing all the hydraulic components in the system and their interconnections. Most hydraulic components can be described by standard symbols. Standard symbols (see chapter 2), should be used wherever possible. They are recommended for use worldwide and are published as an international standard (ISO 1219) and as national standards (such as ANSI Y32.10, DIN 24300, BS 2917, and DS 109.3). Typical circuit diagrams with standard symbols are shown in chapter 3.

Note that the standard symbols describe components in a neutral way by focusing on the type of function and the operational state of each separate component. The symbols do not describe specific makes, actual physical design principles or dimensions.

The circuit diagram must present a clear overview of the system (ref. 20) and should be shown from bottom to top in the direction of the power flow.

> Hydraulic actuators
> Control valves
> Pump
> Fluid reservoir

The position of the component symbols and the connections in a circuit diagram will not normally correspond to their locations in the physical system. Also, detailed pipe connections need not be shown. The circuit diagram is two dimensional, whereas the physical system is three dimensional.

In Fig. 9.6 the design procedure for designing a hydraulic circuit is illustrated for a simple case. The design starts on the basis of a specified function-time diagram (see Fig. 9.6a). Such a diagram specifies the hydraulically controlled machine functions typified by the actuators. In this example two cylinders are required. The diagram in Fig. 9.6a (ref. 20) illustrates the control commands (signals) initiating the actuator movements in the required sequence as well as the direction flow control via a valve Y.

In Fig. 9.6b, c and d the three groups of components–fluid reservoir and pump, control valves and actuators–needed for the design of a hydraulic system performing the specified sequence are indicated in standard symbols.

A circuit diagram of the above example system is shown in Fig. 9.7. Often pipe connections are labelled with standard designations: P = Pump, R = Return or Reservoir, A/B... = Actuator connections, L = Leakage, x,y,z... = Pilot (control) connections. Normally the circuit components, e.g., the directional valves and the cylinder/piston actuators, should be shown in their neutral state (zero position) or in the position that corresponds to the initial position for the circuit sequence.

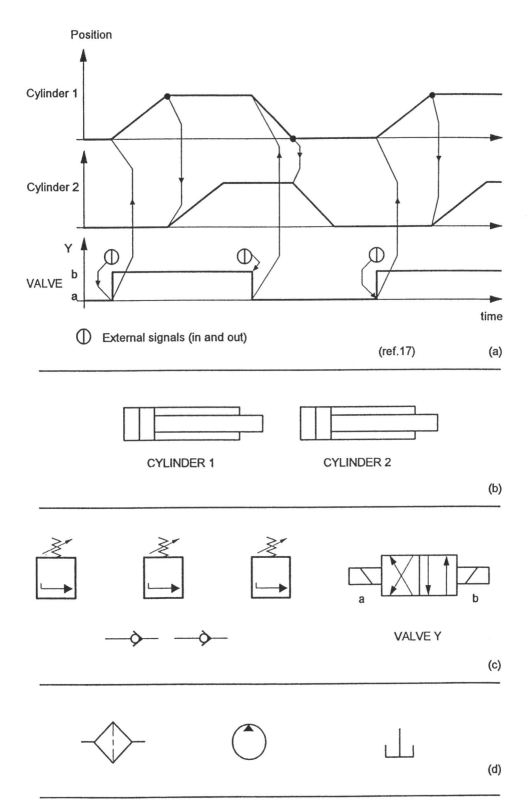

Fig. 9.6 Elements of a circuit design:
(a) required function-time sequence, (b) actuators, (c) valves, (d) power supply

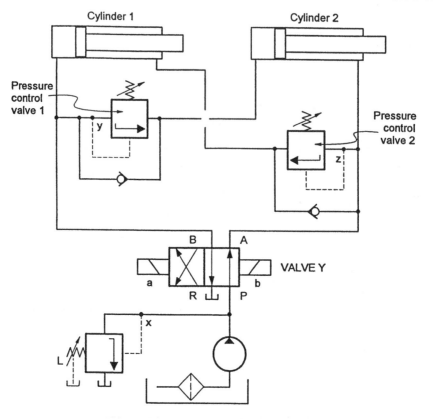

Fig. 9.7 Circuit example (see Fig. 9.6)

When the hydraulic system is complicated, it may be helpful to use a color code for marking corresponding connections on the circuit diagram with the physical pipe connections. The latter can be marked with colored tape or paint. The standard color code is as follows:

Pressure line	red	———
Return line	blue	———
Pilot line, pressure	red	--------
Pilot line, return	blue	--------
Leakage line	blue	--------
Suction line	green	———

In order to ensure a complete documentation of a hydraulic system, the circuit diagram should include a bill of materials identifying all the components contained in the system with regard to number, type, capacity, supplier, material, etc.

As with other technical domains the process of designing more complex hydraulic systems very much relies on the ingenuity and experience of the designer.

In order to avoid misunderstandings, the design of a hydraulic system should be carried out in concert with the machine design project in which the hydraulic control system is going to be integrated. During the design process, changes in the design specifications frequently occur, and the design should document these for reference and accommodate these as quickly as possible.

9.3 Design rules

Choice of hydraulic pressure

When the specifications such as the load loci for all actuators to be used in a hydraulic design have been defined (see Fig. 9.4), the maximum load force and the maximum speed for each individual actuator are specified. Then the maximum hydraulic pressure to be used in the system can be chosen by sizing the displacements of the actuators. Several factors influence this choice, depending on the particular case. In principle a wide range of choice is available, but in practice a compromise optimizing functionality **and** costs has to be made.

The higher the maximum pressure chosen, the smaller the actuators, pumps, valves, and pipe dimensions will be, but the price of many of the components will increase due to higher manufacturing costs attributable to increased precision and tighter tolerances. The lower the maximum pressure chosen, the bigger (and heavier) the components will be, and beyond a certain point this will also lead to higher prices. But also at lower pressures leakage is easier to control, seals become simpler, friction forces due to pressure load are smaller and throttle control in valves (assuming the same maximum actuator speeds) become easier to handle due to higher volume flow rates. Furthermore, at lower pressures, the noise level is lower and there is less vibration.

In practice various suppliers of marketed standard hydraulic components, responding to market needs, have defined certain levels for the maximum working pressure to be used in their components. In Table 9.1 an overview of typical maximum working pressures in various **water** hydraulic applications is listed.

Table 9.1

Typical maximum pressures in present water hydraulic applications

Industrial hydraulics	
Presses	300 bar
Steel plants and rolling mills	200 bar
Industrial production lines and machinery	175 bar
High-pressure cleaning	175 bar
Mining machinery	400 bar
Off-shore equipment	250 bar

Hydraulic pump size

Having specified the maximum speeds and the maximum load forces for each individual actuator from the load loci, as shown in Fig. 9.4, and using the selected maximum working pressure (see the previous section), a designer can derive the displacements and the flow rate requirements for each actuator.

By inspection of the specified working sequence in the function-time diagram (see Fig. 9.5), the coincidence of motions of the various actuators can be identified. Now the maximum total required flow rate to the actuators to be delivered from the pump can be derived. The pump must be rated accordingly so it can deliver this flow rate at the maximum required working pressure, adjusted by a pressure relief valve. Due to the lack of water hydraulic pumps with variable displacement, large power losses may occur during periods of the work cycle when only small flow rates are needed in the system. In such cases the energy loss may be minimized by combining a constant-displacement pump with an accumulator system or by using two or more constant-displacement pumps (see Fig. 3.8).

Example. Dimensioning of a hydraulic pump and a motor

Using the analysis in section 5.2 and pump and motor data specified in sections 5.4(c) and 6.2, and for the sake of simplicity assuming that only one rotary motor is active, the effective (required) maximum power E_{eP} on the pump input shaft can be derived as follows:

$$E_{eP} = \frac{(T_{eM})_{Max} \cdot (n_M)_{Max}}{\eta_{tP} \cdot \eta_{tM}}$$

Having chosen the hydraulic working pressure p_P, one can derive the motor displacement by:

$$D_M = \frac{(T_{eM})_{Max} \cdot 2\pi}{p_P \cdot \eta_{mM}}$$

and the required, rated pump flow rate Q_P is derived by:

$$Q_P = \frac{D_M \cdot (n_M)_{Max}}{\eta_{vP} \cdot \eta_{vM}}$$

and the pump displacement by:

$$D_P = \frac{Q_P}{n_P}$$

The quantities $(T_{eM})_{Max}$ and $(n_M)_{Max}$ are taken from the motor load locus in question (see Fig. 9.8).

The effective motor volume flow rate, Q_{eM}, using the results from section 5.2, can be derived by:

$$Q_{eM} = \frac{D_M \cdot (n_M)_{Max}}{\eta_{vM}}$$

and the effective, maximum output torque $(T_{eM})_{Max}$ by:

$$(T_{eM})_{Max} = \frac{D_M \cdot p_P}{2\pi} \cdot \eta_{mM}$$

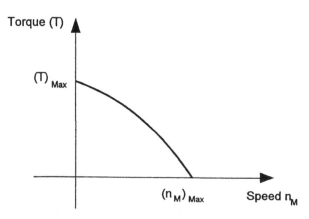

Fig. 9.8 Load locus for a rotary motor

Cylinder dimensions

Considering a double-acting cylinder with single-end rod (see Fig. 6.10e), the theoretical piston force, F_{Cyl}, during full extension of the piston can be computed by:

$$F_{Cyl} = \frac{\pi}{4}(p_1 \cdot D^2 - p_2(D^2 - d^2))$$

where p_1 = pressure at full cross-sectional area
p_2 = pressure at annulus cross-sectional area
D = cylinder full bore diameter
d = piston rod diameter.

The effective piston force, F_{eCyl}, due to friction forces from the seals and the fluid in the cylinder, can be expressed by:

$$F_{eCyl} = \eta_{mCyl} \cdot F_{Cyl}$$

where η_{mCyl} is defined as a mechanical efficiency of the cylinder. An approximated value is $\eta_{mCyl} = 0.9$.

Considering the load loci for the cylinder actuator in question (see Fig. 9.4), the above-computed value of F_{eCyl} must satisfy the inequality:

$$F_{eCyl} = \eta_{mCyl} \cdot \frac{\pi}{4}(p_1 \cdot D^2 - p_2(D^2 - d^2)) > F_{Max}$$

The maximum required theoretical flow rate, Q_{Cyl}, to the cylinder is determined by the maximum required velocity v_{Max} (see Fig. 9.4) of the piston and can be computed by:

$$Q_{Cyl} = \frac{\pi}{4} \cdot D^2 \cdot v_{Max}$$

In case of leakage in the cylinder a volumetric efficiency, $\eta_{v\,Cyl}$, for the cylinder can be defined, and the maximum required effective flow rate, Q_{eCyl}, is expressed by:

$$Q_{eCyl} = \frac{\frac{\pi}{4} \cdot D^2 \cdot v_{Max}}{\eta_{v\,Cyl}}$$

Two nomograms for the computation of (1) piston speed, flow rate and cylinder area and (2) piston force, cylinder area and hydraulic working pressure are shown in Figs. 9.9 and 9.10, respectively.

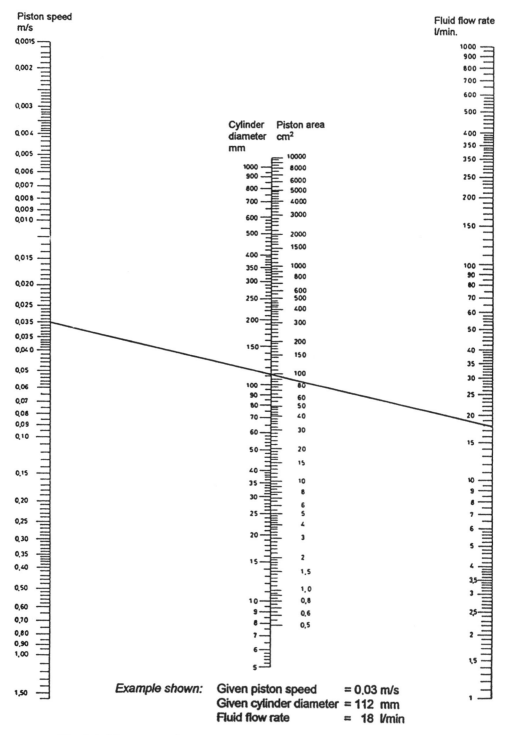

Fig. 9.9 Nomogram for computation of piston speed and fluid flow rate

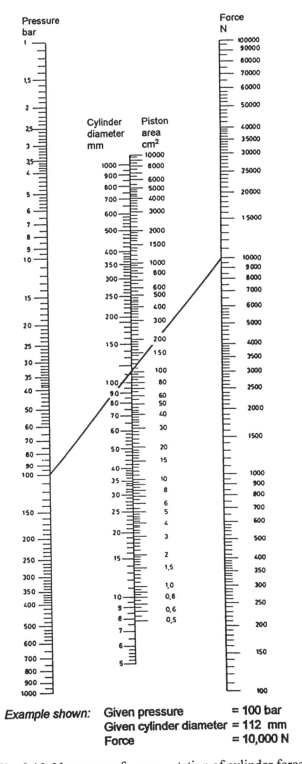

Example shown: Given pressure = 100 bar
Given cylinder diameter = 112 mm
Force = 10,000 N

Fig. 9.10 Nomogram for computation of cylinder force

Pipe line dimensions

Fluid flow through pipes involves corresponding pressure drops across the pipe ends. The higher the flow velocity, the higher the pressure drop. Such pressure drops should be kept as low as possible so that a high power transmission efficiency can be achieved. Therefore, the pipe dimensions selected must ensure acceptable fluid flow velocities.

The low viscosity of water compared with the viscosity of most other hydraulic pressure media justifies the use of higher flow velocities in water hydraulic systems. Two other relationships, however, must be considered.

The relatively high vapor pressure of water at the preferred operational temperature interval, 3°–50°C, increases the risk of cavitation in the suction line of a pump (see Fig. 8.4). The higher the flow velocity in the suction line, the more the absolute pressure will decrease, according to the Bernoulli equation (see section 1.3).

When water flow velocities in the selected dimension of a pipe line are estimated, it should be recognized that **turbulent flow conditions** in water hydraulic systems prevail and the flow rate Q may then be derived by using the following formula (repeated from section 1.3):

$$Q^{1.75} = \frac{D^{4.75}}{0.242 \cdot \mu^{0.25} \cdot \rho^{0.75}} \cdot \frac{p_1 - p_2}{L}$$

As an example to illustrate differences in pressure drops $(p_1 - p_2)$ and in flow rates (Q) between water and mineral oil, the flow through the same pipe line of the two different hydraulic pressure fluids is considered. In both cases turbulent flow conditions are assumed. By using the above formula, ratios for the two cases can be computed: (1) the ratio of the drop in pressure $(p_1 - p_2)$ when the flow rate (Q) is the same and (2) the ratio of the flow rate (Q) when the drop in pressure $(p_1 - p_2)$ is the same. The viscosity of water is assumed to be $\mu_W = 1$ cP and of mineral oil to $\mu_{oil} = 27$ cP. The mass density of water is assumed to be $\rho_W = 1000$ kg/m³, and of mineral oil to be $\rho_{oil} = 900$ kg/m³. The subscript "W" refers to water, and the subscript "Oil" to mineral oil. The ratios are as follows:

$$(1) \quad \frac{\left(\dfrac{p_1 - p_2}{L}\right)_W}{\left(\dfrac{p_1 - p_2}{L}\right)_{Oil}} = \frac{\mu_W^{0.25} \cdot \rho_W^{0.75}}{\mu_{Oil}^{0.25} \cdot \rho_{Oil}^{0.75}} = 0.475$$

$$(2) \quad \frac{Q_W^{1.75}}{Q_{Oil}^{1.75}} = \frac{\mu_{Oil}^{0.25} \cdot \rho_{Oil}^{0.75}}{\mu_W^{0.25} \cdot \rho_W^{0.75}} = 0.106$$

$$\frac{Q_W}{Q_{Oil}} = 1.531$$

From this example it appears that for the same flow rate the pressure drop for the water flow is approximately half that for mineral oil flow (ratio 1). For the same pressure drop along the length L of the pipe line the flow rate (and the flow velocity) for the water flow is approximately 50% higher than the flow rate for the mineral oil flow (ratio 2).

One can conclude that given the same rated flows and pressures, a designer may choose water hydraulic components of smaller dimensions than their oil hydraulic counterparts.

When a fluid flows through a pipe and is suddenly stopped, for example, because a valve at the end of the pipe suddenly closes fast, a large pressure transient may occur. This phenomenon, named **water hammer,** generates a loud noise. The pressure peak in the pressure transient, p_{Max}, may be computed approximately by the following formula (ref. 10):

$$p_{Max} = \rho \cdot c \cdot v_0$$

where

p_{Max} = pressure peak
ρ = mass density of fluid
c = velocity of sound in fluid
v_0 = velocity of fluid prior to valve closure

Introducing typical values for ρ and c in the above equation, the transient pressure peak, p_{Max}, is computed for water and mineral oil, respectively, for a fluid velocity at 5 m/sec in the pipe. The results are shown in Table 9.2 and illustrate the increase in water-hammering effects when using water as the hydraulic pressure medium. Note that the damping out of the pressure transients in water takes a much longer time than in oil because of the relatively low viscosity of water.

Table 9.2

Comparison of water hammer effects

	c (m/sec)	ρ (kg/m³)	v_0 (m/sec)	p_{Max} (bar)
Water	1500	1000	5	75.0
Mineral oil	1300	870	5	56.6

From the above reasons the following fluid flow velocities in water hydraulic pipe connections can be recommended as follows (ref. 11):

Pressure lines :   3–8 m/sec
Return lines :   2–5 m/sec
Suction lines :   0.5–1 m/sec

In Fig. 9.11 a nomogram for computation of pipe diameter, fluid flow velocity and fluid flow rate is given.

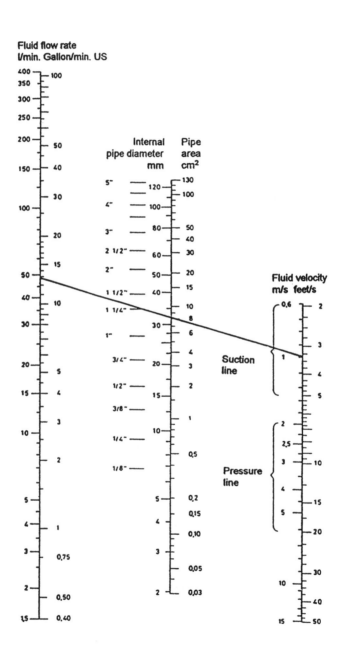

Example shown: Given flow rate   = 50 l/min
                  Given fluid velocity   = 1 m/s
                  Internal pipe diameter   = 32 mm

Fig. 9.11 Nomogram for computation of pipe diameter

Miscellaneous considerations

Various water hydraulic **valves** are described in chapter 7. Valves are purposely inserted restrictions in hydraulic pipes in order to control flow and pressure in the hydraulic system. Engineering data on pressure and flow characteristics of the valves can be found in chapter 7.

In chapter 8 various descriptions and recommendations are given on **power supplies**, **filters** and **accumulators**.

In many hydraulic systems flexible connections are needed. In water hydraulic systems **flexible hoses** may be manufactured from thermoplastic materials (for example, Nessie® hoses) and consist of an inner hose of seamless polyamide plus either one or two layers of fiber reinforcement and an external coating of wear-resistant polyurethane. The hose material may be specifically formulated for using water as the hydraulic pressure medium. The internal hose surface must be resistant to water and must not liberate toxins or other adverse substances into the hydraulic medium or the surrounding environment. Finally, cleaning should be easy and the hose material should be resistant to normal cleaning agents used in the food industry.

For water hydraulic systems **fittings** of stainless steel (AISI 316) or brass (MS 58), for example, Nessie® fittings, are recommended. These fittings are of the JIC 37º flare type. They are of a simple design and easy to assemble (see Fig. 9.12). The sleeve and nut fasten the flared tube and fitting together. The sealing is carried by the 37º surface between the fitting itself and the tube flare. Due to the low viscosity of water, the fitting connections should be mounted even more carefully than in oil hydraulics if leakages are to be avoided.

Seals in water hydraulic systems are normally not more difficult to deal with than seals in oil hydraulic systems. Seal suppliers offer a wide range of seals with acceptable lifetimes for water hydraulic systems and should be consulted for verifying seal compatibility. Due to the low viscosity of water, dynamic seals should be of a self-lubricating type or made of a filled-type seal elastomer. For static seals O-rings of Perbunan, for instance, may be used.

9.4 Maintenance

General

Maintenance can be considered to contain two different concepts: monitoring and preventive maintenance.

Monitoring includes all maintenance tasks that can be carried out during the normal operation of the hydraulic system or during normal servicing.

For any hydraulic system a checklist should be worked out defining all points or tasks of monitoring such as:

- Check operating pressures
- Check pressure-relief valves
- Check fluid level in reservoir
- Check leakages: pipes, fittings, reservoir and components
- Check filter condition indicators
- Check temperature
- Check security precautions
- Check precharge of accumulators
- Check for noise levels (especially pump)

Preventive maintenance or inspection is of utmost importance because a sudden breakdown of a hydraulic system will probably cause a whole automatic production line to stop. The preventive maintenance consists of a preplanned schedule for inspecting/ replacing key components in the system before they break down.

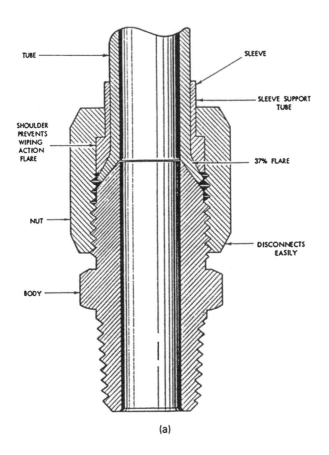

(a)

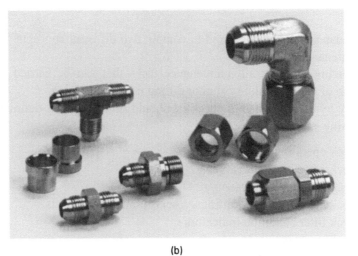

(b)

Fig. 9.12 Fittings of the JIC 37º flare type (Parker Hannifin)
(a) cross-section of fitting (ref. 7); (b) miscellaneous fittings.

The replacement cycles should be defined by operational experience, consulting suppliers and sound engineering judgement.

Typical inspection intervals are:

- Pumps and rotary motors   5,000–10,000 hours
- Cylinder seals   1–5 million strokes
- Directional valves   5,000–10,000 hours
- AC solenoids:
    - no. of operations   1–2 million
- DC solenoids:
    - no. of operations   2–4 million
- Reservoir   1,000–4,000 hours
    - (clean tank and exchange fluid pressure medium).

Water hydraulic systems

To ensure a long service life of trouble-free operation for a water hydraulic system, some practical guidelines should be followed:

1. Oil must be removed from hoses, pipes, and fittings. Note that compressed air may contain oil drops for lubrication!
2. Good mechanical and personal hygiene should be observed by the operational personnel.
3. Components must be rinsed with water in order to remove glycol.
4. Protection caps must be applied for quick connectors.
5. Drain cocks must be placed at strategic locations at the water hydraulic system in order to be able to take bacterial samples.
6. The correct type of materials must be used for fittings, manifolds, pipes, etc. Typical materials are: stainless steel, disinfectant-resistant brass, and plastic.
7. The water temperature should be kept at 40°C for several hours when a new water hydraulic system is started up. It is further recommended that the water be changed frequently during the first operational period. This can be judged by inspection of the state of the water.
8. The water level in the reservoir must be monitored periodically. Water replenishment (amount and time) and water pH and hardness measurements should be carried out periodically and the data recorded for future reference.
9. Microbiological hygiene should likewise be monitored by measuring the amount of microorganisms in the water and recorded for future reference. Two measuring methods are applicable:
    - The "dip slice" method: $>10^3$ bacteria/ml (resolution)
    - The "Petri dish" method (exact measurement): > 1 bacterium/ml (resolution)

    These measurements should be made as quickly as possible after sampling. Measurements should be taken at 21°C and 37°C.

9.5 Applications

The modern revival of water hydraulics for power transmission in industrial systems is based upon a conceptual approach, where a complete range of water hydraulic components, including hydraulic motors, pumps, valves, power packs and accessories, is available on the market to enable the design and making of complete water hydraulic systems. The main benefits of water hydraulics–safety, environmental friendliness and low operational costs–however, will generally be achieved only if all components in the water hydraulic systems are designed and developed to meet all requirements of using water as hydraulic pressure medium. The application of hydraulic components developed for different types of pressure medium in water hydraulic systems may jeopardize the function of the system due to corrosion, increased leakage and lack of lubrication.

In Fig. 9.13 a water hydraulic system based upon the Nessie® product range of water hydraulic components is illustrated. The power supply consists of a pump with constant displacement, reservoir, filters, cooler, pressure-relief valves and pressure gauge. There are 4 actuators: two double-acting cylinders, each with a single-end rod; one unidirectional rotary motor; and one bidirectional rotary motor. The flow to and from the actuators is controlled by directional valves, and the speed is controlled by throttle valves or flow-control valves.

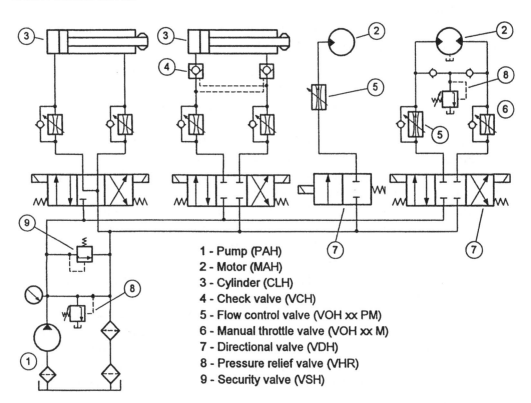

1 - Pump (PAH)
2 - Motor (MAH)
3 - Cylinder (CLH)
4 - Check valve (VCH)
5 - Flow control valve (VOH xx PM)
6 - Manual throttle valve (VOH xx M)
7 - Directional valve (VDH)
8 - Pressure relief valve (VHR)
9 - Security valve (VSH)

Fig. 9.13 A water hydraulic system based upon the Nessie® product range

A conveyor system

The conveyor system shown in Fig. 9.14 illustrates an application of water hydraulics. The system consists of Danfoss Nessie® components and illustrates the versatility of this program. A diagram is shown in Fig. 9.15.

The conveyor is driven by an axial piston–type motor with a displacement of 12.5 cm^3 connected to a planetary gearbox with a 6.3:1 ratio and lubricated with approved grease (FDA). The conveyor may be raised or lowered by means of a water hydraulic cylinder whose effective diameter is 40 mm and whose rod diameter and stroke are 20 mm and 400 mm, respectively. The conveyor water hydraulic circuit incorporates:

- One fixed-displacement axial piston pump
- One fixed-displacement axial piston motor
- One double-acting cylinder with single-end rod
- Three flow-control valves
- Two 4-port, 3-position directional control valves electrically actuated
- One water hydraulic check valve
- One suction and one return filter
- One pressure-relief valve

Fig. 9.14 A conveyor system drive

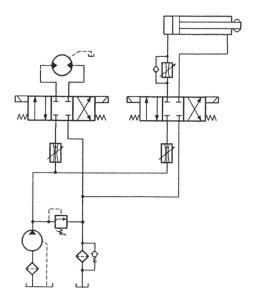

Fig. 9.15 Water hydraulic system for the conveyor in Fig. 9.14

Rib-top saw system

A rib-top saw system is illustrated in Fig. 9.16. Such systems are manually operated and are used for sawing through pork ribs and backs in meat-packing plants. This is a very special operation requiring a saw with a horizontal blade to give the operator the best working position.

The conventional system consists a pneumatic motor with a maximum output of 3.4 kW, operating at speeds up to 4500 rpm and at a noise level of 90 dB(A). The compressed air consumption at maximum output is 58 l/sec at 7 bar. When idling, the consumption is 12 l/sec. When the air decompresses, the formation of ice on the saw motor makes the saw heavy and cold to touch.

(a)

(b)

Fig. 9.16 A rib-top saw:
(a) the saw and its power supply; (b) the saw at a meat-processing line

A new system based upon a water hydraulic power system has been developed. A circuit diagram of the system is shown in Fig. 9.17.

The heart of the saw consists of a lightweight, compact, axial piston water motor, which is ideal for powering hand tools. The motor displacement is 10 cm³. This enables speeds of up to 3000 rpm to be reached, resulting in a blade speed of more than 30 m/sec. The system's working pressure is 80 bar and the flow rate is 28 l/min, giving a motor output torque of 12 Nm and a power consumption of 3.4 kW. The volumetric efficiency of the motor is approximately 95%, resulting in a total system efficiency of nearly 90%. The noise level has been reduced to less than 75 dB(A).

The rib-top saw incorporates a manually controlled speed control valve as well as a built-in safety brake, which is automatically engaged when the operator releases the hand-operated control.

The saw has been designed so that the feed and return of the pressurized water supply may be uncoupled simultaneously via a quick-acting coupling system.

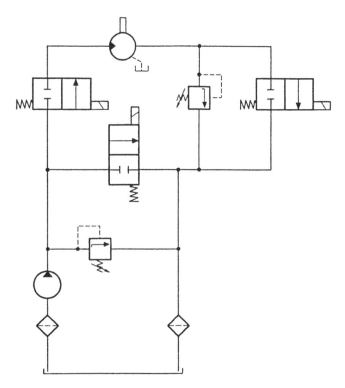

Fig. 9.17 Water hydraulic system for the rib-top saw in Fig. 9.16

A sludge pump drive system

A company produces and distributes drinking water, removing water from a river and processing it to a quality fit for drinking. The process plant includes a sludge pump that removes the thickened sludge from the treated water and pumps it to a filter press for the removal of the remaining water. The sludge pump is built as an ordinary piston pump with a valve system based on check valves.

Previously the drive system for the piston pump included an oil hydraulic cylinder controlled by conventional oil hydraulic spool-type throttle check valves. As can be seen in Fig. 9.18, the use of an oil hydraulic system was not without problems. The continuous operation and the consequent wear on the piston pump seals resulted in a greasy mixture of oil and water. The emulsion threatened to find its way through leaking piston seals and contaminate the drinking water. The result was a shutdown of the plant and a subsequent cleaning process. Because of the lack of a suitable alternative, this latent danger of contamination of the drinking water was accepted, albeit unwillingly.

The drive has now been exchanged with a water hydraulic system with an improved performance: a Nessie® system incorporating a double-acting cylinder with a single-end rod in a differential coupling (see the hydraulic diagram in Fig. 9.19). The coupling ensures the same piston speed in both directions of the hydraulic cylinder and avoids the use of valves for speed adjustment. The water hydraulic drive system also totally eliminates the contamination of the drinking water produced by the process plant because the pressure medium in the hydraulic pump drive is itself pure drinking water!

Fig. 9.18 Previous oil hydraulic drive for sludge pump

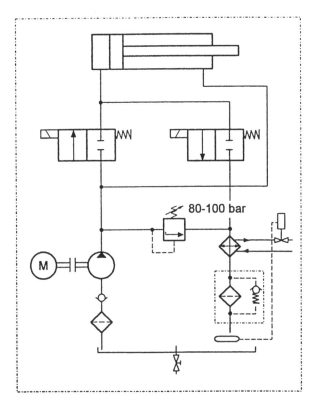

Fig. 9.19 Diagram of water hydraulic drive for sludge pump

The engineering data of the system shown in Fig. 9.19 are listed in Table 9.3

Table 9.3

Engineering data of the system in Fig. 9.19

Hydraulic cylinder	diameter Ø 80/56 mm
	stoke 225 mm
Pump displacement	25 cm^3 per rev.
Pump speed	1460 rpm @ 200 bar
Pump vol. efficiency	0.93
Pump output flow rate	33.9 l/min
System pressure	100 bar
Pump input power	6 kW

10. CONCLUSION: OUTLOOK FOR WATER HYDRAULIC TECHNOLOGY

Up to the beginning of this century the only hydraulic pressure medium used was water or water emulsions. The pressures (<100 bar) and pump speeds (<100 rpm) were modest, although a few experimental hydraulic systems used considerably higher pressures. In the beginning of this century mineral oil was introduced as the pressure medium in hydraulic systems, and the installed hydraulic power drives grew in size. The working pressure increased (up to 500 bar), and high-speed pumps (up to 4000 rpm) came into use, so the power of installed units increased by 1–2 orders of magnitude. Although this development was initiated by the replacement of water with mineral oil, it was also strengthened by the application of new types of bearings and improved innovative designs.

Through the first several decades of this century the development of water hydraulics stagnated, and for many years it was used primarily for hydraulic presses in heavy industries, such as in forges, shipyards, steel mills, and the mining industry. Since the early eighties, however, new application areas such as in automobile, food-processing, and nuclear industries, have generated an increasing interest in water hydraulics.

The starting point for this change has been a growing mistrust in using mineral oil in hydraulics, considering the increasing attention industry must pay to the following:

- improved safety at the working place
- improved environmental protection

These requirements have added new criteria for the selection of pressure media for hydraulic systems beyond the traditional technical ones, which are criteria for **viscosity**, **lubrication** and **wear**, and **ageing properties** (ref. 16). The new criteria may be classified as follows:

- non-flammability
- product compatibility
- biological degradability

Good **non-flammability** means that the hydraulic fluid does not pose any fire risk. Good **product compatibility** means that the produced product will not be contaminated by leaks from the hydraulic fluid when a hydraulic-powered production system is used. Good **biological degradability** means that no long-lasting contamination of floor, soil, or ground-water will occur if the hydraulic system should leak.

Taking all the above criteria into consideration makes water, and especially clear water (tap water), an interesting and attractive medium for hydraulics. However, the physical properties of water differ significantly from those of mineral oil. Because of this, industry has, until now, widely believed that high-pressure water hydraulics for power control would be too difficult and costly to use for general automatic machine control.

A changeover of the pressure medium from mineral oil to water can in general not be done just by modifying oil hydraulic components. The significant differences in physical properties of the two types of pressure media require a shift in **design paradigm**. Water hydraulic components and systems are of unique design, fulfilling their specific requirements. The most significant differences in physical properties between water and mineral oil are summarized in the following section.

Significant differences in physical properties between water and mineral oil in hydraulics

Viscosity

Viscosity is the physical property underscoring the most dramatic difference between water and mineral oil. Viscosity values referring for typical operational conditions are as follows: for **water**, 1 cS @ atmospheric pressure and 20°C (see Fig. 4.14); for **mineral oil**, 30 cS @ atmospheric pressure and 55°C (see Fig. 4.15). Put another way, the viscosity of water is 30 times smaller than the viscosity of mineral oil.

Also, viscosity as a function of temperature and pressure is quite different for water and for mineral oil: for **water** the viscosity varies by a factor of ~3 in the temperature range 3–50°C and the pressure range 1–1000 bar (see Fig. 4.14); for **mineral oil** the viscosity will varies by a factor of ~150 in the temperature range 20–70°C and the pressure range 1–1000 bar (see Fig. 4.15).

Vapor pressure

The vapor pressure is also quite different for water and mineral oil: for **water** the vapor pressure @ 50°C is ~0.12 bar (see Fig. 4.16); for a **mineral oil** the vapor pressure @ 50°C is ~$1.0 \cdot 10^{-8}$ bar (see Fig. 4.16). In other words, the vapor pressure at 50°C is 10^7 times higher for water than for mineral oil.

Air solubility

As described in chapter 4 (see Fig. 4.8) the amount of air dissolved under the same conditions in a saturated water-air solution and in a saturated mineral oil-air solution is approximately 2% vol. and 10% vol., respectively. This means that compared with water, mineral oil may contain five times the amount of air in solution.

Mass density

The mass density of **water** is ~1000 kg/m³ (see Fig. 4.5) and of **mineral oil** is ~900 kg/m³ (see Fig. 4.4), which means that the mass density of water is ~10% higher than the mass density of mineral oil.

Speed of sound

The speed of sound in **water** is ~1480 m/sec and in **mineral oil** is ~1300 m/sec @ 20°C (see Fig. 4.16), which means that the speed of sound is ~14% higher in water than in mineral oil.

Compression modulus

The compression modulus (see Fig. 4.16) for **water** is $2.4 \cdot 10^4$ bar and for **mineral oil** is $1.6 \cdot 10^4$ bar, which means that the compression modulus for water is 50% higher than for mineral oil.

The higher compression modulus and the higher speed of sound for water lead to faster pressure time responses and to smaller pressure decompression shocks because of water's lesser amount of potential energy in fluid pressure volumes, such as in cylinder chambers in hydraulic presses.

Thermal conductivity and specific heat

The thermal conductivity (see Fig. 4.16) of water is 4–5 times that of **mineral** oil. This means that water systems tend to require less cooling capacity. The specific heat of **water** is two times that of **mineral oil**, so water has double the ability to absorb heat.

Summary of special challenges in water hydraulics

The most important problems to be **overcome** in the design and development of **water** hydraulic components are caused by:

- Aggressive corrosion by water of the materials used for the hydraulic components and their connections.
- Much faster and more severe erosion and abrasion of the materials of the components due to cavitation from the high vapor pressure, low viscosity and high velocity of water (ref. 1).
- Greater wear of and high friction between the sliding parts in components (such as journal bearing and piston/cylinder motions) due to the low viscosity of water.
- High leakages in clearances due to low viscosity of water.
- Pressure peaks and noise from water-hammering effects.
- Limited range for the operational temperature of the water.

The problems may be overcome by (1) selecting the appropriate types of materials such as stainless steel, bronze and brass and in some cases such materials as polymers, anodized alumina, and ceramics (2) adapting new and flexible layouts and design principles for water hydraulic components such as in bearings and in clearances in throttling valves, (3) optimizing the internal geometrical layout of the components in order to achieve the desired flow and pressure states in accordance with the physical properties of water and ensuring the correct functioning of leakage passages, clearances and damping chambers, and (4) designing the correct dimensioning of a heat exchanger (water/cooler) in the power pack so the operational temperature of water remains in the interval 3–50°C.

Outlook

As described in the preceding chapters, it can be recognized that several different water hydraulic components and pieces of equipment are available on the market today. Pumps, valves, rotary as well as linear motors, filters, accumulators, power supplies, and so on are uniquely designed and developed as standard components for operating entirely on pure water (tap water). The components are available for the design and building of practical, complete water hydraulic power control systems and may be used, e.g., in food-stuff production machinery in the meatpacking and dairy industries. Many more new applications are on their way.

Also, it is expected that several new water hydraulic components such as proportional valves, servovalves and pumps with variable displacement will be introduced onto the market. In this way the range of standard water hydraulic components will be enhanced to offer more possibilities for making even better and more competitive water hydraulic systems in the future.

The performance and operating life of pure tap water hydraulic components described in this text are quite comparable to their oil hydraulic counterparts. Water hydraulic systems and their components, furthermore, are completely fireproof, offer a high level of hygiene for the products made using systems and pose no risk of contamination of the internal and external environment from possible fluid leakage or waste.

In Fig. 10.1 an overview of existing and future application areas for water hydraulics is indicated.

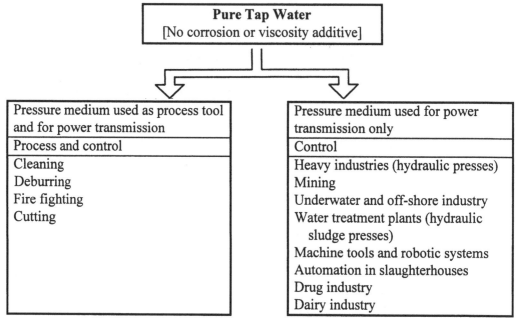

Fig. 10.1 Application areas for pure tap water hydraulics

As shown, water hydraulic systems can be used for two types of applications: (1) where the water is used for processing and for power control and (2) where the water is used for power transmission control only.

The modern evolution of water hydraulics is only a little more than a decade old. In spite of this, mature product programs are already being, marketed and complete water hydraulic systems can be built from standard components. In Appendix E a list of manufacturers making water hydraulic components/systems is included. Already today in several areas pure tap water hydraulic control systems are often a cost-effective solution–in some cases the only solution.

With continued advances in water hydraulic technology driven by environmental concerns and market demands, water hydraulic components and systems will offer functional performance, efficiency and operating service life competitive with other fluid power systems.

REFERENCES

(1) Backè, W. and J. Berger: "Kavitationserosion bei HFA-Flüssigkeiten," O+P "Ölhydraulik und Pneumatik" 28 (1984) No. 5, pp. 288–295.

(2) Blackburn J.F., G. Reethof, and J.L. Shearer: *Fluid Power Control*, MIT Press and John Wiley & Sons, Inc. New York and London, 1960.

(3) McCloy, D. and H.R. Martin: *The Control of Fluid Power*, Longman Group, London, 1973.

(4) Currie, J.A.: "The development of Raw Water Hydraulics," First Bath International Fluid Power Workshop, University of Bath, England, 8th September, 1988.

(5) *Design Engineers Handbook*. Parker-Hannifin Corporation, Cleveland, Ohio, USA.

(6) Fitch, E.C. and J.B. Surjaatmadja: *Introduction to Fluid Logic*. McGraw-Hill, 1978.

(7) Gleen, R.E. and J.E. Blinn: *Mobile Hydraulic Testing*, American Technical Society, Chicago, 1970.

(8) Kirk-Othmer: *Encyclopedia of Chemical Technology*, Third Edition, Volume 24, John Wiley & Sons, New York · Chichester · Brisbane · Toronto · Singapore, 1984.

(9) Lambeck, Raymond P.: *Hydraulic Pumps and Motors: Selection and Application for Hydraulic Power Control Systems*, Marcel Dekker, Inc., New York and Basel, 1983.

(10) Merritt, Herbert E.: *Hydraulic Control Systems*, John Wiley & Sons, New York, London, Sydney, 1967.

(11) Müller, Ernst: *Hydraulische Pressen und Druckflüssigkeitsanlagen*, Springer-Verlag, Berlin/Göttingen/ Heidelberg, 1955.

(12) O+P–Report 1992, "Geräte der Druckwasserhydraulik," p. 127.

(13) Oehler, Gerhard: *Hydraulic Presses*, Edward Arnold, 1968.

(14) Panzer, Paul and Gerhard Beitler: *Arbeitsbuch der Ölhydraulik*, 2nd edition, Krausskopf-Verlag, 1969.

(15) Pinches, Michael J. and John G. Ashby: *Power Hydraulics*, Prentice Hall, Englewood Cliffs, New Jersey, 1988.

(16) Rinck, Stefan: "Die Wasserhydraulik entwickelt sich und findet ihre Anwendungs-bereiche," O+P "Ölhydraulik und Pneumatik" 36 (1992), No.1, pp. 25–32.

(17) Schlösser, W.M.J.: "Analogien bei Antrieben," VDI-Z 115 (1973) No. 7, May, pp. 560–568.

(18) *Technical Data on Shell Tellus Oil*, Published by Shell International Petroleum Company, Ltd., London, 1963.

(19) Trostmann, E.: *Hydraulic Control,* Lecture notes, rep. S87.51, IFS, Technical University of Denmark (in Danish), 1987.

(20) *VDI–Richtlinien für Funktionsdiagramme von Arbeitsmaschinen und Fertigungs-anlagen*, VDI 3260, July, 1977, Beuth Verlag GmbH, Berlin and Cologne.

(21) Correspondence with Hauhinco Maschinenfabrik, Mr. Brian Hollingworth, D-45549 Sprockhövel, Germany, and Hytar Oy Water Hydraulics, Mr. Olle Pohls, SF-33101 Tampere, Finland.

APPENDIX A

Conversion of units

The measurable quantities dealt with in hydraulics are primarily pressure, force, velocity, volume flow rate (in short referred to as flow), density, viscosity and temperature.

These variables are related through relationships expressed as equations derived from physical laws or definitions. Each variable contains one or more of the basic dimensions length (L), mass (M), force (F), time (T) and temperature (K). A set of units must be defined for these dimensions.

Equations expressing relations among physical variables must be dimensionally **homogeneous**. This means that each term in the equations must be expressed in the same dimensions. To avoid difficulties, the units of mass and force cannot be selected independently.

Newton's second law illustrates that force and mass are **not** independent dimensions. Force F is proportional to the product of mass m and acceleration a as given by

$$F = m \cdot a$$

Actual choice of basic dimensions may be somewhat arbitrary. In Table 1.A below the dimensions mass (M), length (L) and time (T) are used.

Table 1.A

Physical variable	Dimensions	SI units	English units
Length	L	m (meter)	ft (foot)
Mass	M	kg (kilogram)	lbm (pound mass)
Time	T	sec	sec
Velocity	LT^{-1}	m/sec	ft/sec
Acceleration	LT^{-2}	m/sec^2	ft/sec^2
Force	MLT^{-2}	Newton	lbf (pound force)
Energy. work	ML^2T^{-2}	Newton m (Joule)	ft lbf
Power	ML^2T^{-3}	Joule/sec (Watt)	ft lbf/sec
Density	ML^{-3}	kg/m^3	lbm/ft^3
Angular velocity	T^{-1}	rad/sec	rad/sec
Angular acceleration	T^{-2}	rad/sec^2	rad/sec^2
Torque	ML^2T^{-2}	Newton m	ft lbf
Moment of inertia	ML^2	kg m^2	lbm/ft^2
Pressure	$ML^{-1}T^{-2}$	Newton/m^2	lbf/in^2
Viscosity, dynamic	$ML^{-1}T^{-1}$	Newton sec/m^2	lbf sec/ft^2
Viscosity, kinematic	L^2T^{-1}	m^2/sec	ft/sec^2

In Table 2.A currently used conversion factors between different units are listed.

Table 2.A

Length	1 meter (m) = 10^2 centimeter (cm) = 10^3 millimeter (mm) = 10^6 micron (μ) = 10^9 millimicron (mμ) = 10^{10} ångström (Å) 1 inch (in) = 2.540 cm 1 foot (ft) = 30.48 cm 1 mile (mi) = 1.609 km 1 mil = 0.001 in = 0.0254 mm
Area	1 square meter (m^2) = 10.76 ft^2 1 square foot (ft^2) = 929 cm^2
Volume	1 liter = 1000 ml = 1000 cm^3 = 61.03 in^3 = 1.057 quart (qt) 1 US gallon (gal) = 4 quarts = 231 in^3 = 3.785 l 1 British gallon (gal) = 277.4 in^3 = 4.545 l
Mass	1 kilogram (kg) = 10^3 gram (g) = 10^6 mg = 2.205 lb$_m$ = 0.0685 slug
Force	1 Newton (N) = 0.1020 kilopond (kp) = 10^5 dynes = 0.2248 lbf 1 metric ton = 2205 lbf 1 US short ton = 2000 lbf 1 US long ton = 2240 lbf
Energy. Work	1 Newton meter (Nm) = 1 Joule = 0.2389 cal = 10^7 ergs = 0.7376 ft lbf = $9.481 \cdot 10^{-4}$ Btu
Power	1 Watt (W) = 1 Joule/sec = 0.2389 cal/sec = 10^7 ergs/sec 1 kilowatt (kW) = 1.341 horsepower (Hp) = 737.6 ft lbf/sec = 0.9483 Btu/sec 1 Hp = 0.7457 kW 1 kW = 238.9 cal/sec
Pressure	1 N/m^2 = 1 Pascal (Pa) = 10^{-5} bar = $9.869 \cdot 10^{-6}$ (physical) atmosphere (atm) 1 technical atmosphere (at) = 1 kp/cm^2 = 0.981 bar 1 physical atmosphere (atm) = 76 cm Hg (mercury) = 10.333 m H$_2$O = 14.70 lbf/in^2 = 14.70 pounds per square inch (psi)

Table 2.A (continued)

Temperature	Conversion	Celsius ($t_{°C}$) to Fahrenheit ($t_{°F}$)
		$t_{°C} = (t_{°F} - 32.0) / 1.8$
	Conversion	Celsius ($t_{°C}$) to Kelvin ($t_{°K}$)
		$t_{°C} = t_{°K} - 273.15$
Viscosity, dynamic		1 N sec/m² = 10^5 Dyn sec/m²
		= 10 Poise (P) = 10^3 Centipoise (cP)
		1 kp sec/m² = 98.1 P
		1 Reyn = lbf sec/in² = $6.895 \cdot 10^6$ cP
		1 Reyn = 10^6 µR
Viscosity, kinematic		1 m²/sec = 10^4 cm²/sec
		= 10^4 Stoke (S) = 10^6 Centistoke (cS)
		1 cS = 0.01 S = 1 mm²/sec

APPENDIX B

Conversion of units for kinematic viscosity (ref. 18)

Centi-stokes	Saybolt Seconds at			Redwood Seconds at			Engler
	39.8 °C	54.4 °C	99 °C	21.1 °C	60 °C	93.3 °C	Degrees at all temps.
1.0		29.3			26.7		1.00
1.50		31.3			28.4		1.07
2.0	32.6	32.7	32.8	30.2	31.0	31.2	1.14
3.0	36.0	36.1	36.3	32.7	33.5	33.7	1.22
4.0	39.1	39.2	39.4	35.3	36.0	36.3	1.31
5.0	42.3	42.4	42.6	37.9	38.5	38.9	1.40
6.0	45.5	45.6	45.8	40.5	41.0	41.5	1.48
7.0	48.7	48.8	49.0	43.2	43.7	44.2	1.56
8.0	52.0	52.1	52.4	46.0	46.4	46.9	1.63
9.0	55.4	55.5	55.8	48.9	49.1	49.7	1.73
10.0	58.8	58.9	59.2	51.7	52.0	52.6	1.84
11.0	62.3	62.4	62.7	54.8	55.0	55.6	1.93
12.0	65.9	66.0	66.4	57.9	58.1	58.8	2.02
14.0	73.4	73.5	73.9	64.4	64.6	65.3	2.22
16.0	81.1	81.3	81.7	71.0	71.4	72.2	2.43
18.0	89.2	89.4	89.8	77.9	78.5	79.4	2.64
20.0	97.5	97.7	98.2	85.0	85.8	86.9	2.87
22.0	106.0	106.2	106.7	92.4	93.3	94.5	3.10
24.0	114.6	114.8	115.4	99.9	100.9	102.2	3.34
26.0	123.3	123.5	124.2	107.5	108.6	110.0	3.58
28.0	132.1	132.4	133.0	115.3	116.5	118.0	3.82
30.0	140.9	141.2	141.9	123.1	124.4	126.0	4.07
32.0	149.7	150.0	150.8	131.0	132.3	134.1	4.32
34.0	158.7	159.0	159.8	138.9	140.2	142.2	4.57
36.0	167.7	168.0	168.9	146.9	148.2	150.3	4.83
38.0	176.7	177.0	177.9	155.0	156.2	158.3	5.08
40.0	185.7	186.0	187.0	163.0	164.3	166.7	5.34
42.0	194.7	195.2	196.1	171.0	172.3	175.0	5.59
44.0	203.8	204.2	205.2	179.1	180.4	183.3	5.83
46.0	213.0	213.4	214.5	187.1	188.5	191.7	6.11
48.0	222.2	222.6	223.8	195.2	196.6	200.0	6.37
50.0	231.4	231.8	233.0	203.3	204.7	208.3	6.63
60.0	277.4	277.9	279.3	243.5	245.3	250.0	7.90
70.0	323.4	324.0	325.7	283.9	286.0	291.7	9.21
80.0	369.6	370.3	372.2	323.9	326.6	333.4	10.53
90.0	415.8	416.6	418.7	364.4	367.4	376.0	11.84
100.0	462.0	462.9	465.2	404.9	408.2	416.7	13.16

APPENDIX C

Water hardness specifications:
Conversion of German, English, French, and U.S.
water hardness units

To / From	Earth-alkali ions (mmol/1)	German degree (^{o}d)	English degree (^{o}e)	French degree (^{o}f)	ppm $CaCO_3$ (U.S.)
Earth-alkali ions (mmol/1)	1.00	5.60	7.02	10.00	100.0
German degree	0.18	1.00	1.25	1.78	17.8
English degree	0.14	0.798	1.00	1.43	14.3
French degree	0.10	0.560	0.702	1.00	10.0
ppm $CaCO_3$ (U.S.)	0.01	0.056	0.0702	0.10	1.00

APPENDIX D

ISO/CETOP Symbols

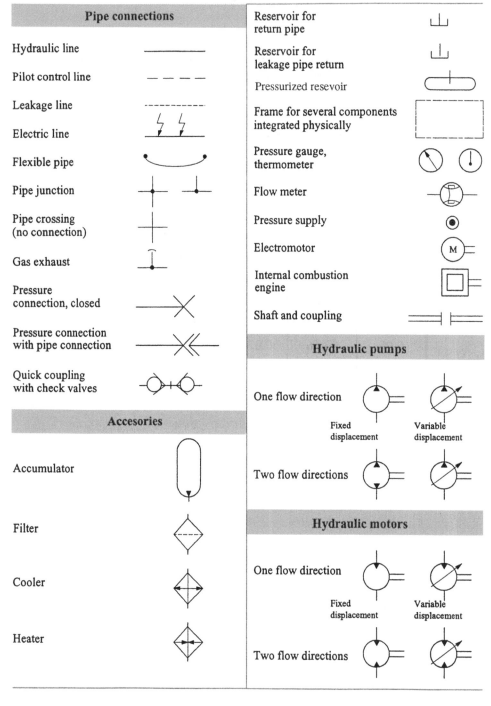

Pipe connections	
Hydraulic line	
Pilot control line	
Leakage line	
Electric line	
Flexible pipe	
Pipe junction	
Pipe crossing (no connection)	
Gas exhaust	
Pressure connection, closed	
Pressure connection with pipe connection	
Quick coupling with check valves	

Accesories	
Accumulator	
Filter	
Cooler	
Heater	

Reservoir for return pipe	
Reservoir for leakage pipe return	
Pressurized resevoir	
Frame for several components integrated physically	
Pressure gauge, thermometer	
Flow meter	
Pressure supply	
Electromotor	
Internal combustion engine	
Shaft and coupling	

Hydraulic pumps	
One flow direction	Fixed displacement / Variable displacement
Two flow directions	

Hydraulic motors	
One flow direction	Fixed displacement / Variable displacement
Two flow directions	

This compilation of graphical symbols is in accordance with DIN/ISO 1219

ISO/CETOP SYMBOLS (continued)

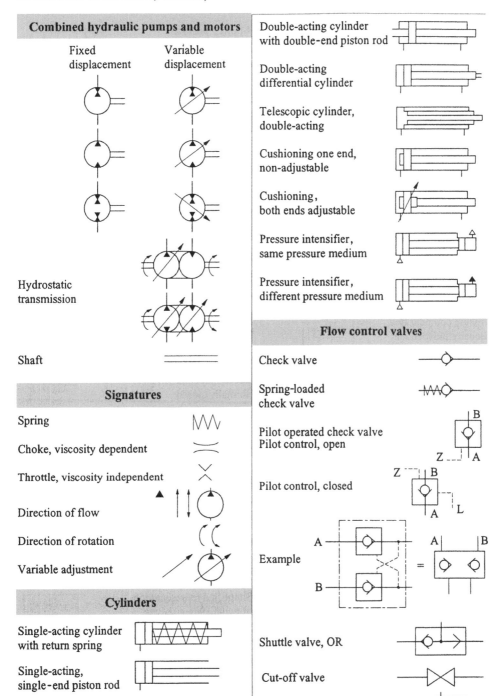

Combined hydraulic pumps and motors	
Fixed displacement	Variable displacement

Hydrostatic transmission

Shaft

Signatures

Spring

Choke, viscosity dependent

Throttle, viscosity independent

Direction of flow

Direction of rotation

Variable adjustment

Cylinders

Single-acting cylinder with return spring

Single-acting, single-end piston rod

Double-acting, single-end piston rod

Double-acting cylinder with double-end piston rod

Double-acting differential cylinder

Telescopic cylinder, double-acting

Cushioning one end, non-adjustable

Cushioning, both ends adjustable

Pressure intensifier, same pressure medium

Pressure intensifier, different pressure medium

Flow control valves

Check valve

Spring-loaded check valve

Pilot operated check valve
Pilot control, open

Pilot control, closed

Example

Shuttle valve, OR

Cut-off valve

Double cut-off valve, AND

ISO/CETOP SYMBOLS (continued)

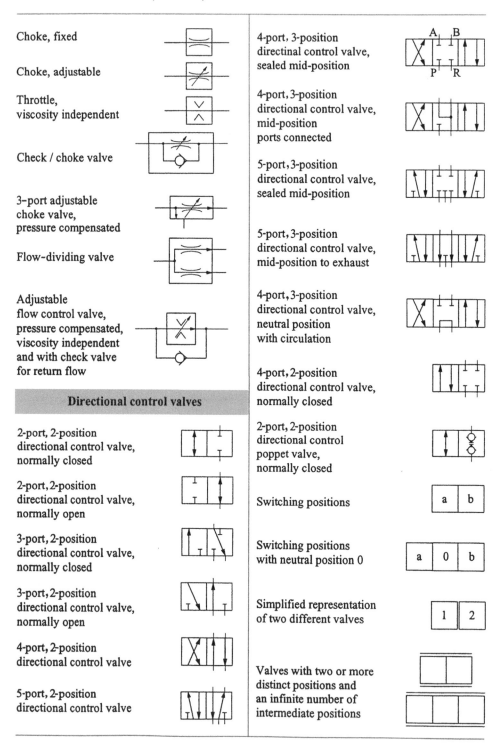

Choke, fixed

Choke, adjustable

Throttle,
viscosity independent

Check / choke valve

3–port adjustable
choke valve,
pressure compensated

Flow–dividing valve

Adjustable
flow control valve,
pressure compensated,
viscosity independent
and with check valve
for return flow

Directional control valves

2-port, 2-position
directional control valve,
normally closed

2-port, 2-position
directional control valve,
normally open

3-port, 2-position
directional control valve,
normally closed

3-port, 2-position
directional control valve,
normally open

4-port, 2-position
directional control valve

5-port, 2-position
directional control valve

4-port, 3-position
directinal control valve,
sealed mid-position

4-port, 3-position
directional control valve,
mid-position
ports connected

5-port, 3-position
directional control valve,
sealed mid-position

5-port, 3-position
directional control valve,
mid-position to exhaust

4-port, 3-position
directional control valve,
neutral position
with circulation

4-port, 2-position
directional control valve,
normally closed

2-port, 2-position
directional control
poppet valve,
normally closed

Switching positions

Switching positions
with neutral position 0

Simplified representation
of two different valves

Valves with two or more
distinct positions and
an infinite number of
intermediate positions

ISO/CETOP SYMBOLS (continued)

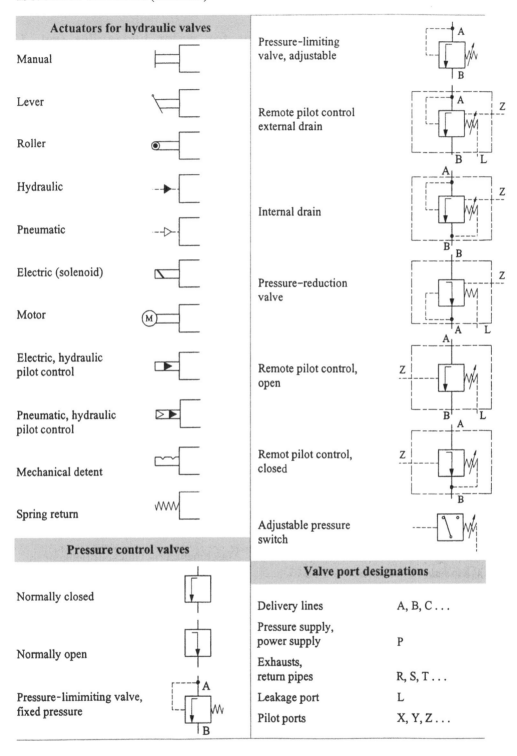

Actuators for hydraulic valves	
Manual	
Lever	
Roller	
Hydraulic	
Pneumatic	
Electric (solenoid)	
Motor	
Electric, hydraulic pilot control	
Pneumatic, hydraulic pilot control	
Mechanical detent	
Spring return	

Pressure control valves	
Normally closed	
Normally open	
Pressure-limimiting valve, fixed pressure	

Pressure-limiting valve, adjustable	
Remote pilot control external drain	
Internal drain	
Pressure–reduction valve	
Remote pilot control, open	
Remot pilot control, closed	
Adjustable pressure switch	

Valve port designations	
Delivery lines	A, B, C . . .
Pressure supply, power supply	P
Exhausts, return pipes	R, S, T . . .
Leakage port	L
Pilot ports	X, Y, Z . . .

APPENDIX E

Manufacturers of components/systems for water hydraulics (ref. 12)

Dr. Breit GmbH
Carl-Zeiss-Straße 25
D-42566 Heiligenhaus
Germany
+49 (0 20 56) 51 69
+49 (0 20 56) 5 74 84

Danfoss A/S
DK-6430 Nordborg
Denmark
+45 74 88 22 22
+45 74 49 09 49

Elwood Corp.
195-A W. Ryan Rd.
Oak Creek, WI 53154
USA
+1 414-764-7500
+1 414-764-4298

Fenner Fluid Power Ltd
Ashton Road
GB-RM3 8UA Romford
United Kingdom
+44 (01708) 343851
+44 (01708) 373993

Hansa - TMP
Via M.L. King 6
I-41100 Modena
Italy
+39 059 253415
+39 059 253299

Hauhinco Maschinenfabrik G. Hausherr,
Jochums GmbH & Co. KG
Beisenbruchstraße 10
D-45538 Sprockhövel
Germany
+49 (0 23 24) 70 50
+49 (0 23 24) 7 05-222

Heyer-HPE Systeme GmbH
Niebuhrstr. 59
D-46049 Oberhausen
Germany
+49 (02 08) 20 30 80
+49 (02 08) 20 10 15

Hunt Valve Co., Inc.
1915-T E. State St.
Salem, OH 44460
USA
+1 216-337-9535
+1 216-337-3754

HydroWatt GmbH
Nördliche Uferstraße 4
D-76189 Karlsruhe
Germany
+49 (0721) 55 72 53
+49 (0721) 59 36 70

Fritz Lotterer GmbH & Co. KG
Am Elsashufer 3
D-72563 Bad Urach
Germany
+49 (0 71 25) 82 95
+49 (0 71 25) 7 08 22

Maschinenfabrik Glückauf Beukenberg
GmbH & Co. KG
Wilhenminenstraße 120
D-45801 Gelsenkirchen
Germany
+49 (02 09) 4 09 90
+49 (02 09) 40 99-189

Modul Hydraulik Weber & Weber GnbR
Hauptplatz 23
A-2474 Gattendorf
Austria
+43 (0 21 42) 64 26
+43 (0 21 42) 64 34

Oilgear GmbH
Im Gotthelf 8-10
D-65786 Hattersheim
Germany
+49 (0 61 45) 37 70
+49 (0 61 45) 3 07 70

Olaer-Industries S.A.
B.P. 7
F-92704 Colombes Cedex
France
+33 1-41 19 17 00
+33 1-41 19 17 20

Parker Hannifin
17325 Euclid Ave.
Cleveland, OH 44112
+1 800-272-7537
+1 216-486-0618

SAIP
Via Adige 4
I-20090 Opera
Italy
+39 02 57603913
+39 02 57604752

Schmidt, Kranz & Co. GmbH
Hauptstraße 123
D-42531 Velbert
Germany
+49 (0 20 52) 88 80
+49 (0 20 52) 8 88 44

Tiefenbach GmbH
Nierenhofer Straße 68
D-45243 Essen
Germany
+49 (02 01) 4 86 30
+49 (02 01) 4 86 31 58

URACA Pumpenfabrik Urach GmbH &
Co. KG
Sirchingen Straße 5-7
D-72563 Bad Urach
Germany
+49 (0 71 25) 1 20
+49 (0 71 25) 12-202

Walter Voss Metallwarenfabrik GmbH u.
Co. KG
Stadeweg 16
D-58102 Hagen
Germany
+49 (0 23 34) 24 35
+49 (0 23 34) 4 20 21

Wepuko Hydraulik GmbH & Co.
Max-Eyth-Straße 31
D-72542 Metzingen
Germany
+49 (0 71 23) 1 80 50
+49 (0 71 23) 4 12 31

Werner & Pfleiderer GmbH
Theodorstraße 10
D-70452 Stuttgart
Germany
+49 (07 11) 89 70
+49 (0711) 8 97-3999

INDEX